International Conference on Sustainable Development and Green Technology (SDGT 2017)

Journal of Physics: Conference Series Volume 989

Chiayi, Taiwan
24 – 26 November 2017

ISBN: 978-1-5108-6162-6
ISSN: 1742-6588

Printed from e-media with permission by:

Curran Associates, Inc.
57 Morehouse Lane
Red Hook, NY 12571

Some format issues inherent in the e-media version may also appear in this print version.

Printed by Curran Associates, Inc. (2018)

For permission requests, please contact the Institute of Physics at the address below.

Institute of Physics
Dirac House, Temple Back
Bristol BS1 6BE UK

Phone: 44 1 17 929 7481
Fax: 44 1 17 920 0979

techtracking@iop.org

Additional copies of this publication are available from:

Curran Associates, Inc.
57 Morehouse Lane
Red Hook, NY 12571 USA
Phone: 845-758-0400
Fax: 845-758-2633
Email: curran@proceedings.com
Web: www.proceedings.com

Table of contents

Volume 989

International Conference on Sustainable Development and Green Technology (SDGT 2017)

24–26 November 2017, Chiayi, Taiwan

Accepted papers received: 15 March 2018
Published online: 11 April 2018

Preface

Papers

Energy and Chemical Engineering

Urban Planning and Sustainable Development

Product Design

IOP Conf. Series: Journal of Physics: Conf. Series **989** (2018) 011001 doi:10.1088/1742-6596/989/1/011001

PREFACE

It is our great pleasure to welcome you to 2017 International Conference on Sustainable Development and Green Technology (SDGT 2017) which has been held Nanhua University, Chiayi, Taiwan during November 24-26, 2017. SDGT 2017 is dedicated to issues related to Sustainable Development and Green Technology.

The major goal and feature of the conference is to bring academic scientists, engineers, industry researchers together to exchange and share their experiences and research results, and discuss the practical challenges encountered and the solutions adopted. Professors from Taiwan and Japan are invited to deliver keynote speeches regarding latest information in their respective expertise areas. It will be a golden opportunity for the students, researchers and engineers to interact with the experts and specialists to get their advice or consultation on technical matters, sales and marketing strategies.

This proceedings present a selection from papers submitted to the conference from universities, research institutes and industries. All of the papers were subjected to peer-review by conference committee members and international reviewers. The papers selected depended on their quality and their relevancy to the conference. The volume tends to present to the readers the recent advances in the field of Sustainable Development and Green Technology and various related areas, such as Sustainability Science, Innovative Technology and Sustainability Engineering, Textiles and Clothing Sustainability, Materials Processing Technology and Materials Science, etc..

We would like to thank all the authors who have contributed to this volume and also to the organizing committee, reviewers, speakers, chairpersons, sponsors and all the conference participants for their support to SDGT 2017.

Assoc. Prof. Yaoming Hong
Nanhua University, Taiwan
December 20, 2017

IOP Conf. Series: Journal of Physics: Conf. Series **989** (2018) 011001 doi:10.1088/1742-6596/989/1/011001

Statement of Peer Review

All papers published in this volume of Journal of Physics: Conference Series (JPCS). have been peer reviewed through processes administered by the proceedings Editors. Reviews were conducted by expert referees to the professional and scientific standards expected of a proceedings journal published by IOP Publishing.

IOP Conf. Series: Journal of Physics: Conf. Series **989** (2018) 011002 doi:10.1088/1742-6596/989/1/011002

Organizing Committee

Conference Co-Chairs

Prof. Tsong-Ming Lin, Nanhua University, Taiwan

Prof. Wen-Pei Sung, college of Humanities and Creativities, National Chin-Yi University of Technology, Taiwan

Prof. Shih-Shiung Chen, College of Science and Technology, Nanhua University, Taiwan

Program Chairs

Assoc. Prof. Yao-Ming Hong, Nanhua University, Dalin Township, Chiayi County, Taiwan

Organizing Chair

Ms. Yang Dan, Hong Kong Chemical, Biological & Environmental Engineering Society, Hong Kong

Publication Chair

Dr. Jinyu Chen, Hunan University, China

Ms. Ivy Hou, Chemistry and Materials Society, HKCBEES, Hong Kong

Technical Program Committee

Prof. Awf Al-Kassir, University of Extremadura, Spain

Prof. Jérôme Chenal, Swiss Federal Institute of Technology in Lausanne, Switzerland

Prof. Huan-Lin Luo, I-Shou University, Taiwan

Prof. Chih-Huang Weng, I-Shou University, Taiwan

Prof. Wu, Edward Ming-Yang,I-Shou University, Taiwan

Assoc. Prof. Serdar Celik, Southern Illinois University Edwardsville, USA

Assoc. Prof. Chien-Sen Liao, I Shou University, Taiwan

Assoc. Prof. Ming-Young Jan, I-Shou University, Taiwan

Assoc. Prof. Fares A. Almoman, Qatar University, Qatar

Dr. Yang Du, Illinois State Geological Survey, USA

Dr. Sayed M. Bateni, University of Hawaii, USA

Dr. Grzegorz Lisak, Nanyang Technological University, Singapore

Dr. Lin Zhang, City University of Hong Kong, Hong Kong

Dr. Chunhua Liu, City University of Hong Kong, Hong Kong

IOP Conf. Series: Journal of Physics: Conf. Series **989** (2018) 011003 doi:10.1088/1742-6596/989/1/011003

Peer review statement

All papers published in this volume of *Journal of Physics: Conference Series* have been peer reviewed through processes administered by the proceedings Editors. Reviews were conducted by expert referees to the professional and scientific standards expected of a proceedings journal published by IOP Publishing.

Comparison of Thermal Performances between Low Porosity Perforate Plate and Flat Plate Solar Air Collector

Hoy-Yen Chan[1], A A Vinson[1], S S S Baljit[1], and M H Ruslan[1]

[1] Solar Energy Research Insititute, the National University of Malaysia (UKM), 43600 Bangi, Selangor, Malaysia
E-mail: hoyyen.chan@ukm.edu.my

Abstract. Flat plate solar air collector is the most common collector design, which is relatively simpler to fabricate and lower cost. In the present study, perforated plate solar collector was developed to improve the system thermal performance. A glazed perforated plate of 6mm holes diameter with square geometry was designed and installed as the absorber of the collector. The influences of solar radiation intensity and mass flow rate on the thermal performance were investigated. The perforated collector was compared with the flat plate solar collector under the same operating conditions. The highest values of thermal efficiency in this study for the perforated plate (PP) and the flat plate (FP) solar collectors were 59% and 36% respectively, at solar radiation intensity of 846 Wm^{-2} and mass flow rate of 0.02 kgs^{-1}. Furthermore, PP collector gave better thermal performance compared to FP collector; and compared to previous studies, the present perforated design was compatible with the flat plate with double pass designs.

1. Introduction

The over-exploit of fossil fuel based energy resources such as crude oil, natural gas, coal and nuclear energy inevitably led to global energy crisis. Considering this, a great research endeavor has been committed to develop renewable and sustainable energy technologies. Among all of the available renewable energies, solar energy is one of the best alternatives owing to its source longevity and especially in recent years, solar based applications have been attracting immense research interests. One of the useful solar applications is solar air heater which harvests solar energy and transforms it into thermal energy. Furthermore, solar air heater is one of the most cost-effective systems that can be applied in both industry and domestic sectors, such as drying agricultural products and space heating [1–3].

There are various types of solar air heaters and the most conventional design for space heating belongs to flat plate (FP) solar air collector which exhibits advantages such as design simplicity and low fabrication cost [4]. FP solar collector consists of: glass, absorber plate and solar absorber plate mounted on insulated back surface, as shown in figure 1a. A number of investigations have been carried out to study and improve the performance of FP solar air collectors in the past [5-9]. Apparently, the thermal performance of FP solar collector in still inadequate due to the significant heat loss and high thermal resistance which in turn, decreases the heat transfer capability between absorber and working fluid [10]. In view of this, various enhancement techniques have been reported and discussed to address the shortcomings of FP solar collector [11-15], e.g. glazed, unglazed, single pass, double pass, perforated solar air collector, etc. There are two known techniques of perforated solar air collector, namely transpired solar collector and impinging jet solar collector.

A simple impinging jet collector model was first developed by Choudhury and Garg [16] and Rask et al. [17]. From the studies, they found an increase in efficiency compared to flat plate solar collector between 10% and 20% depending on the collector configuration and other parameters. Beluko et.al [12] modeled an unglazed impinging jet solar collector and found an increase of efficiency by 21%. There also a study found that jet impingement solar collector can transfer thermal energy with a heat transfer efficiency three times higher than that of FP system [18]. Along with it, there also many studies of transpired solar collector have been reported [19-20]. Despite that, from all the previous studies it is clear that the collector performance of both techniques is affected on many different parameters, such as porosity, geometry of the holes, hole spacing, solar intensity, mass flow rate, surface coating, the collector configurations, etc [21-25]. These indicates that it is a feasible approach to enhance the heat transfer coefficient between the working fluid and plate surface, which will then reduce heat losses and increase the thermal efficiency of the system [26].

In present work, a low porosity perforated plate (PP) was used as the absorber of the solar collector which consists of: glass, absorber plate, perforated plate, and back insulation. The objective of this research is to investigate and study the thermal performance of a glazed perforated plate compared to the flat plate (FP) collector. Our results demonstrated that the present perforated design gave better thermal efficiency compared to the flat plate and it was compatible with the flat plate with double pass designs.

2. Experimental setup and procedure

Solar collector was constructed and indoor experiments were carried out under indoor conditions by using solar simulator as shown in figure 1. The collector was located approximately 1 m bellow the solar simulator. The collector was built with several slots to have several plate distances (X) where the top slot used to place a standard sheet tempered glass as the cover. The outer side of the collector was fully insulated and the back of the collector was insulated with polyurethane slab to prevent the heat loss from the back of the collector. The aluminum perforated and absorber plates were fabricated and painted with black colour as shown in figure 2, and the configuration is as shown in figure 3. Pyranometer CMD3 was used to determine the intensity of the solar simulator. K-type thermocouples and an AM-4214 anemometer were used to measure the temperature and the air flow rate. The specification of the solar collector and the operating condition were shown in tables 1 and 2.

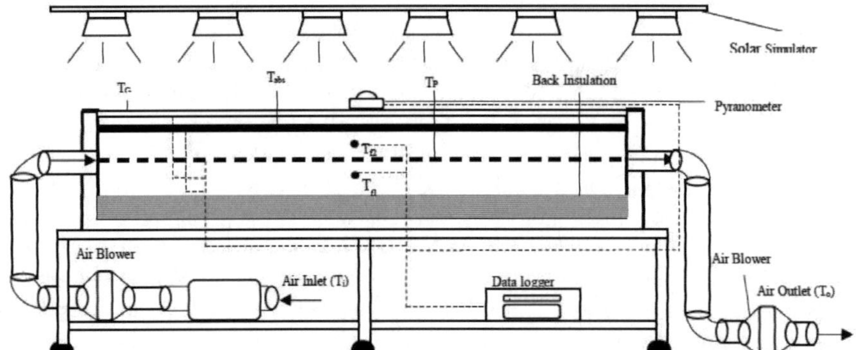

Figure 1. Experimental set up for PP and FP solar collectors.

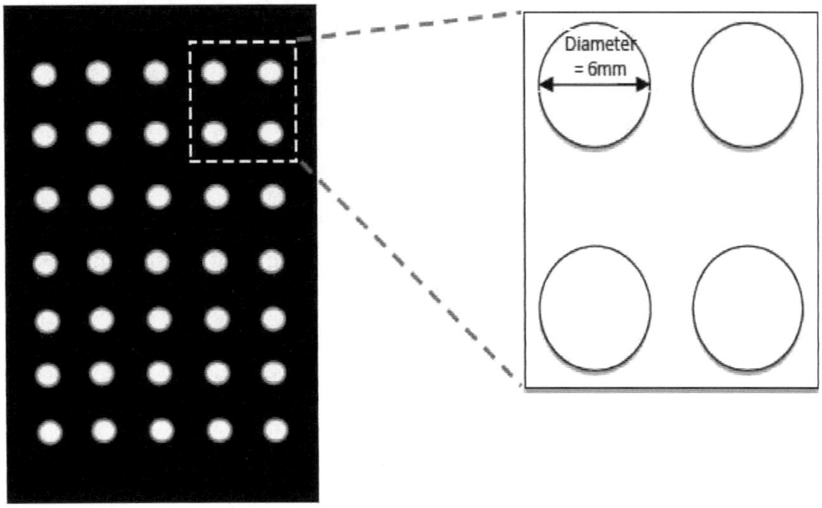

Figure 2. Perforated plate design.

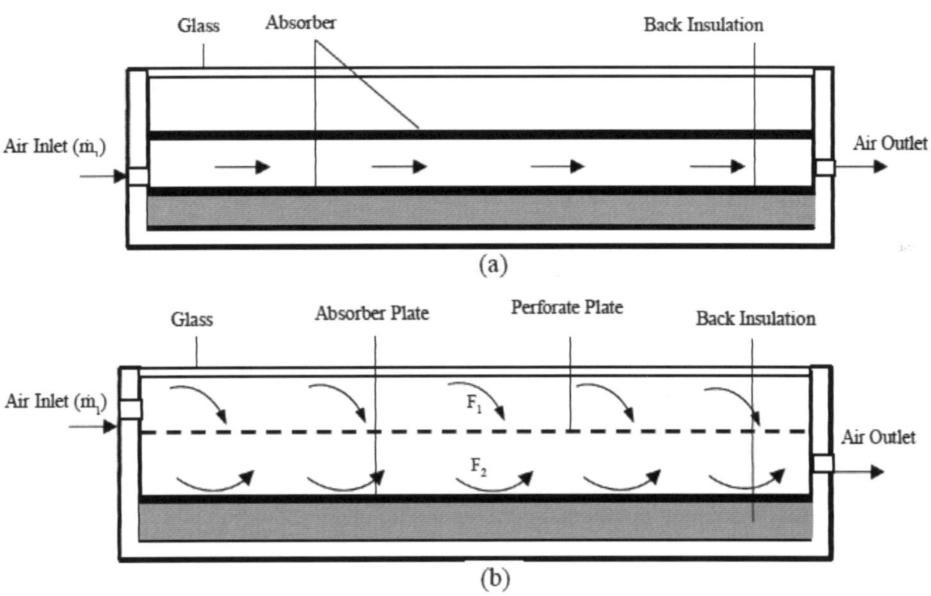

Figure 3. Schematic of glazed a) flat plate and b) perforated solar collectors.

Table 1. Dimensions of the perforated solar collector components

Components	Dimensions
Collector frame	100.5 cm x 69.5 cm x 34 cm

Perforated plate	99.5 cm x 68.5 cm x 0.2 cm
Hole diameter	6 mm
Thickness of insulation	7.7 cm
Thickness of the tempered glass	6 mm
Conductivity of the polyurethane slab	0.007 $Wm^{-2}K^{-1}$
Porosity of the perforated plate	0.97%

Table 2. Operating conditions of the experiment

Parameters	Properties
Solar intensity	350 Wm^{-2} to 1050 Wm^{-2}
Mass flow rate	006 kgs^{-1} to 0.02 kgs^{-1}
Inlet temperature	30°C

3. Data analysis

The following assumptions were used in this study analysis:

(i) the systems were assumed in steady state

(ii) the heat transfer from the sides and bottom of the collector were neglected.

Equation (1) shows the equation of an average thermal efficiency of the solar air collector system [7].

$$\eta = \left[\dot{m}c_p(T_o - T_i)\right](IA_c)^{-1} \quad (1)$$

whereby c_p is the specific heat of air [J kg^{-1} K^{-1}]; $\dot{m}$ is the air mass flow rate [kgs^{-1}]; T_o and T_i are the inlet and outlet air temperature respectively; I is the solar radiation intensity [Wm^{-2}]; and A_C is the area of the collector.

4. Results and discussions

The performance of both PP and FP solar collector were evaluated through the values of temperature difference between the inlet air and out air (ΔT), and the thermal efficiency (η) of the collectors. The influences of solar intensity (I) and mass flow rate ($\dot{m}$) were investigated.

4.1 Effects of solar radiation intensity

The effects of solar radiation intensity on the thermal performance of PP and FP solar *collectors were illustrated in f*igure 4. The values of temperature difference are increasing with solar intensity, which the highest values of temperature difference at 1059Wm^{-2} for PP and FP collectors are about 31°C and 22°C, respectively. While the effect of solar radiation intensity on thermal efficiency is relatively small compared to temperature difference. The highest and lowest thermal efficiency in this study range from 42 to 44% for PP collector, and 30 to 32% for the FP collector. This indicates that in terms of efficiency, solar radiation intensity is not a critical factor, however the impact would be more on the air temperature, which higher radiation intensity gives higher outlet temperature and hence the temperature difference. In addition, results also found that at the same solar radiation intensity, the perforated design has higher thermal performance compared to the flat plate collector.

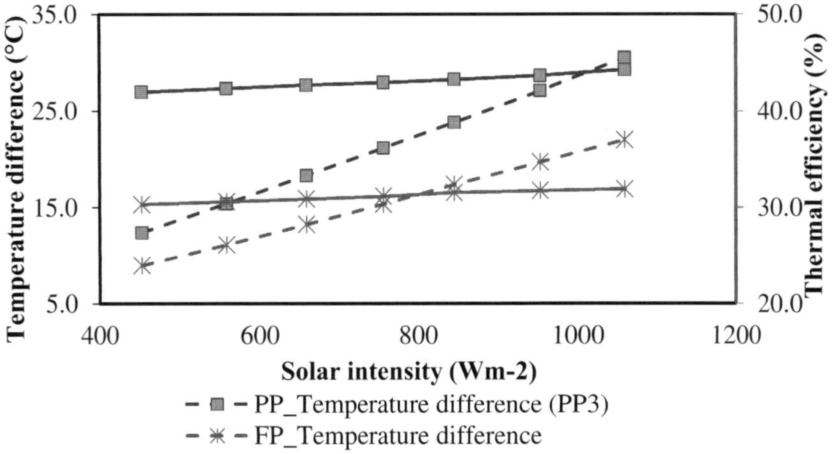

Figure 4. Effect of solar intensity on the thermal performance of PP and FP solar collectors at mass flow rate of 0.0104 kgs^{-1}.

4.2 Effects of mass flow rate

*f*igure 5 shows the effects of mass flow rate on the thermal performance of PP and FP solar collectors. The temperature difference is decreasing while the thermal efficiency is increasing with the mass flow rate. At the highest thermal efficiency value of 59%, the temperature difference value is 17°C for the PP collector, with outlet air temperature of 48°C. The PP collector also shown a better thermal performance compared to a FP collector, whereby at the highest thermal efficiency value for FP is only about 36%, with the temperature difference value of 10oC and outlet air temperature of 42oC.

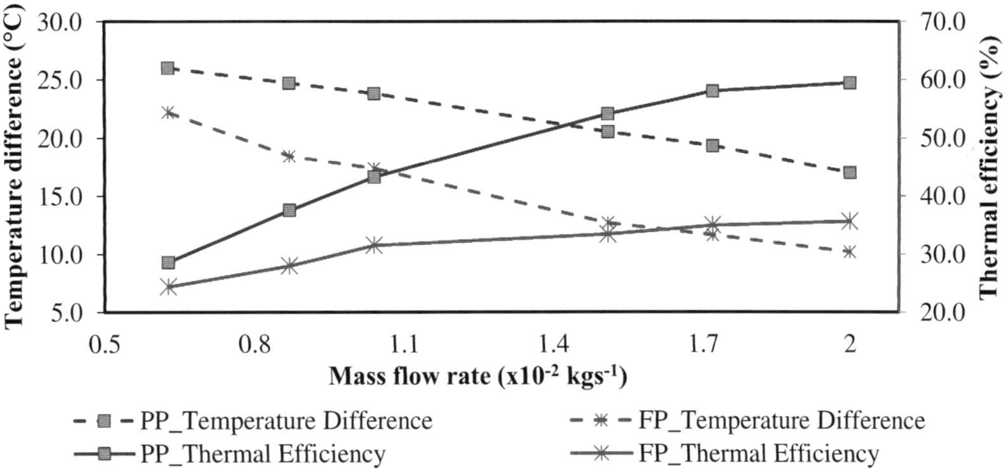

Figure 5. Effects of mass flow rate on the thermal performance of PP and FP solar collectors at solar radiation intensity of 846 Wm^{-2}.

This is due to the fact that the perforated design is able to enhance the heat transfer between the air and the absorber (PP). The heated air was drawn through the holes of the plate and this has minimized the convention heat loss between the glass and PP, and further drawing the heat from the PP. Moreover, the enhanced heat transfer rate through PP design has also reduce the surface temperature of the PP and hence, reduce the radiation heat loss. Apart from that, results also showed that mass flow rate is a key factor in determining the system thermal performance, both the temperature difference and efficiency. Higher mass flow rate gives lower value of outlet temperature. Thus, depend on the applications, the required outlet temperature of the system could be adjusted by controlling the mass flow rates.

4.3 Comparisons with previous studies

Table 3 shows the comparisons of the values of thermal efficiency of solar collectors between present work and other flat plate solar collectors. The comparisons were compiled according to mass flow rate at 0.02 kgs^{-1}. Results showed that the present design of perforated plate gave better thermal efficiency compared to the flat plate with single and double passes, and also better than the vertical transpired solar collectors. However, it is lower than flat plate collector with double pass airflow that consisted matrix layer.

Table 3. Comparison of thermal efficiency of the present study with other flat plate solar collector at mass flow rate of 0.02kg/s

Solar collector type	Thermal efficiency	Reference
Present Glazed Perforated Solar	59%	Present Study
Flat Plate	36%	Present Study
Flat Plate (double pass without matrix)	56%	[27]
Flat Plate (double pass with matrix)	63%	[27]
Glazed Transpired Solar Collector (Vertical)	38%	[26]

5. Conclusion

In this study, the thermal performance of PP and FP solar collectors were investigated. The experiments were carried out in a laboratory under the operating conditions of solar radiation intensity ranged from 350 Wm^{-2} to 1050 Wm^{-2} and mass flow rate ranged from 0.006 kgs^{-1} to 0.02 kgs^{-1}. Present study has found that the thermal efficiency of both PP and FP solar collectors were increasing with solar radiation intensity and mass flow rates. The highest and lowest thermal efficiency under different intensities in this study range from 42 to 44% for PP collector, and 30 to 32% for the FP collector. Whereas, under different mass flow rates, the highest thermal efficiency value for PP and FP was measured 59% and 36%, respectively. While the temperature difference between the inlet and outlet air were increasing with solar radiation intensity and decreasing with mass airflow rate. However, relatively mass flow was the key factor that would influence the system performance and outlet air temperature. Furthermore, PP collector gave better thermal performance compared to FP collector.

6. References

[1] Salunkhe P B and Jaya Krishna D 2017 *J. Energy Storage* **12** 243–60
[2] Al-damook A and Khalil W H 2017 *Renew. Energy* **112** 498–509
[3] Elmegaard B, Ommen T S, Markussen M and Iversen J 2016 *Energy Build.* **124** 255–64
[4] Ho C D, Yeh C W and Hsieh S M 2005 *Renew. Energy* **30** 1601–21
[5] Zhu T, Diao Y, Zhao Y and Ma C 2017 *(MHPA)* vol **118** (Elsevier Ltd)
[6] Jouybari H J, Saedodin S, Zamzamian A, Nimvari M E and Wongwises S 2017 *Renew. Energy*
[7] Hussain S and Harrison S J 2017 *Appl. Therm. Eng.* **123** 940–52
[8] Verma S K, Tiwari A K and Chauhan D S 2017 *Energy Convers. Manag.* **134** 103–15

[9] Bhowmik H and Amin R 2017 *Energy Reports* **3** 119–23
[10] Choudhury C, Chauhan P M and Garg H P 1995 *Renew. Energy* **6** 739–49
[11] Chabane F, Moummi N, Benramache S, Bensahal D and Belahssen O 2013 *Eng. J.* **17** 43–55
[12] Belusko M, Saman W and Bruno F 2008 *Sol. Energy* **82** 389–98
[13] Fudholi A, Sopian K, Ruslan M H and Othman M Y 2013 *Energy Convers. Manag.* **76** 8–19
[14] Krishnananth S S and Kalidasa Murugavel K 2012 *J. King Saud Univ. - Eng. Sci.* **25** 135–40
[15] Ho C D, Hsiao C F, Chang H and Tien Y E 2017 *Energy Procedia* **105** 28–34
[16] Choudhury C, Garg H P and Khas H 1991 **46** 199–209
[17] Rask D R and Mueller L J 1977 *Honeywell, Inc., Minneapolis, MN. Energy Resour. Center.* **1**
[18] Zuckerman N and Lior N 2006 Volume **39** 565–631
[19] Croitoru C V., Nastase I, Bode F I and Meslem A 2016 T *Sol. Energy* **131** 21–9
[20] Hall R and Blower J 2016 *Energy Procedia* **91** 56–63
[21] Navaneethakrishnan P, Krishnakumar K and John A K 2016 *Appl. Therm. Eng.* **106** 432–7
[22] Attalla M and Salem M 2013 *Appl. Therm. Eng.* **51** 723–33
[23] El-Sebaii a. a. and Al-Snani H 2010 *Energy* **35** 1820–8
[24] El-Sebaii a. a. and Al-Snani H 2010 *Energy* **35** 1820–8
[25] Goodro M, Park J, Ligrani P, Fox M and Moon H K 2008 *Int. J. Heat Mass Transf.* **51** 6243–53
[26] Zheng W, Li B, Zhang H, You S, Li Y and Ye T 2016 *Energy* **109** 781
[27] Ramani B M, Gupta A and Kumar R 2010 *Sol. Energy* **84** 1929–37

Acknowledgement
We would like to thank the Universiti Kebangsaan Malaysia for the research grant of GUP-035-2015 in this work.

IOP Conf. Series: Journal of Physics: Conf. Series **989** (2018) 012002 doi:10.1088/1742-6596/989/1/012002

No Cost – Low Cost Compressed Air System Optimization in Industry

A Dharma[1], N Budiarsa[2], N Watiniasih[3], and N G Antara[2]

[1] Electrical Engineering Department, University of Udayana, Bali, Indonesia
[2] Machine Engineering Department, University of Udayana, Bali, Indonesia
[3] Biological Science and Mathematics Department, University of Udayana, Bali, Indonesia
E-mail: agd2bali@gmail.com

Abstract. Energy conservation is a systematic, integrated of effort, in order to preserve energy sources and improve energy utilization efficiency. Utilization of energy in efficient manner without reducing the energy usage it must. Energy conservation efforts are applied at all stages of utilization, from utilization of energy resources to final, using efficient technology, and cultivating an energy-efficient lifestyle. The most common way is to promote energy efficiency in the industry on end use and overcome barriers to achieve such efficiency by using system energy optimization programs. The facts show that energy saving efforts in the process usually only focus on replacing tools and not an overall system improvement effort. In this research, a framework of sustainable energy reduction work in companies that have or have not implemented energy management system (EnMS) will be conducted a systematic technical approach in evaluating accurately a compressed-air system and potential optimization through observation, measurement and verification environmental conditions and processes, then processing the physical quantities of systems such as air flow, pressure and electrical power energy at any given time measured using comparative analysis methods in this industry, to provide the potential savings of energy saving is greater than the component approach, with no cost to the lowest cost (no cost - low cost). The process of evaluating energy utilization and energy saving opportunities will provide recommendations for increasing efficiency in the industry and reducing CO^2 emissions and improving environmental quality.

1. Introduction

Energy conservation is the use of energy efficiently and rationally without reducing the energy usage that is really necessary. Energy conservation efforts are implemented at all stages of utilization, from utilization of energy resources to ultimate utilization, using efficient technology, and cultivating an energy-efficient lifestyle. There are three basic principles in the first energy conservation: eliminating energy waste (prevention); secondly, reducing energy losses (recovery) and the last is increasing the efficiency of energy utilization (innovation).

Industrial growth contributes vital to Indonesia's economic growth, but on the other hand the industry is also the largest energy user and also makes the industry sector has a significant greenhouse gas emission. Increased greenhouse gas emissions from burning fossil fuels and rising world fuel prices pose a serious threat to the environmental sustainability and economic sustainability of Indonesia. The results of a survey conducted by the United Nations Industrial Organization (UNIDO) in 200 Industries in 2009 showed that in general the process involving compressors as well as other equipment such as boilers and pumps is a significant point of energy users. The facts show that energy saving efforts in the process usually only focus on replacing tools and not an overall system improvement effort. The potential of the team involved in this research has been through various stages of education and training conducted by the Ministry of ESDM through the Directorate General of EBTKE, so it is expected that the research can run smoothly.

In this research a systematic technical approach in evaluating accurately a compressed-air system and its optimization potential through observation of environmental conditions and processes, then processing the physical quantities of systems such as flow (air flow), air pressure (pressure) and electric power unity work time (energy) measured using the comparative analysis method in the beverage industry as a case study, in the hope that it can provide greater energy saving potential compared to the component approach in no-cost to low cost cost). This study can also provide a more comprehensive picture of the potential of compressed air system optimization (CASO) as a basis for better action planning steps, as well as to provide better performance and reliability. Energy conservation in addition to the savings gained from energy efficiency the most important thing is also to reduce CO^2 emissions in industrial processes [1].

2. Compressed air system optimization in industry

The program of forming a compressed air system in a country outside Indonesia is a best practice program and is a key component of one of the energy efficiency programs in the industry, initially supported by New Zealand government ministries and agencies. But in this global context it is not a new initiative because the program already exists and works in Europe and the United States, but in each case it affects after ten years and subsequently has been done with limited long-term improvements in overall energy efficiency [2].

Compressed air system installed almost 70% in all plants and not infrequently these systems pay between 20-30% of the total electricity consumption of the company. A review of the system and operational conditions will often result in one or more compressors that are closed or replaced with a more economical type of operation, as it will increase the overall efficiency of the system. To optimize the compressed air system, it is important to thoroughly investigate all equipment systems as well as their respective components. This means that in addition to examining the individual parts of the system, it is also necessary to analyze the inlet system side and discharge and interaction between them [3].

For this research, relevant data must be measured and then combined to yield an analysis of the energy-consumption patterns, under normal operational conditions, of the plant concerned, a sample of Compressed air system optimisation (CASO) in industry was taken in Bali Island. It represented the system that used in industry to produce product that can be finally distribute to the consumer. The type of this system was beverage industry with the following data that related to compressed air system and system operating costs consecutively. It has shown at table 1 and table 2 respectively.

Table 1. The compressed air systems on site

System description	No of compressors	No of Running Compressors	Total rated output, SCFM	Total rated power, kW	Treatment system/dew point
Compressor Room for Production Plant	3	2	661.32	130	Refrigerant

Note: Compressor number 2 and 3 (35 kW) were not used due to out of service.

Table 2. Systems operating costs

Annual generating cost	*12,429.72*	*USD*
	164,010,155	*IDR*
Mean demand used to calculate cost	*230*	*SCFM*
Mean power consumption	*43*	*kW*

Specific power consumption	*18.87*	*kW/100cfm*
Annual energy consumption	*162,480*	*kWh*
CO_2 equivalent	*128,360*	$kgCO_{2eq}$
Unit cost	*0.187*	*$/kWh*
Operating hours per annum	*3,744* *12x6x52*	*Hrs*

3. Finding, discussion and recommendation

The plant air system

The sample plant of this research actually has 1 compressor GA-55 which was in good condition and 2 compressors GA-22 were out of service and need to be repaired. The compressor used as on/off load compressor has no VSD system. The single stage, oil injected, air cooled, GA-55 rotary screw machine distributed air in series to *2000 l* air receiver tank, then it entered it refrigerant air dryer (FD-220 and FD-60) with pre and after filtration and finally is distributed to the production line. Both dryers were piped in *2"* galvanised pipework. The capacity of compressors were *55 kW* and *22 kW* respectively. The measurement has been done in Compressor GA-55, then we get some data like: system pressure, air flow and electrical power consumption.

Figure 3.1. Shows a simplified layout of equipment in compressor house. The equipment installed in compressor room, such as: Compressors, Air Receiver Tank, Air Dryers.

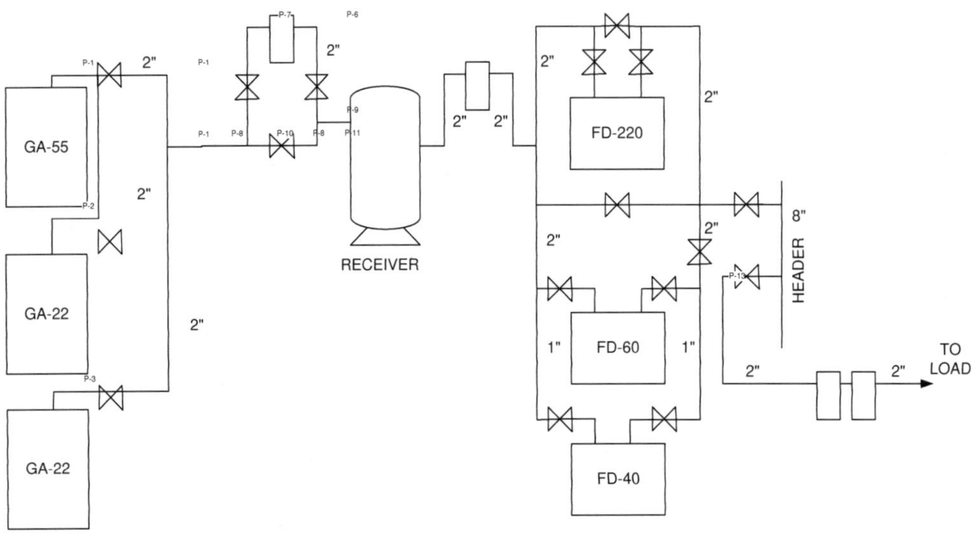

Figure 3.1. Simplified layout of the equipment in compressor house.

(a) *System* Demand

Some measurements was taken during this research such as, pressure system along in wet and dry pipe at a certain point in the compressed air system, air flow, voltage and current being used by motor when load or unload. All of those data's is store to data logger. System demand curve performance of air flow and pressure drop during full load and no load condition as shown in figure 3.2. In this case the factory runs over a 12 hours period from around 06.30 am to 22.30 pm. The first and last 2 hours of this period are devoted into cleaning period. The average production demand is *230scfm*, this drops to around *75scfm* when cleaning. Out of production/cleaning times the system is depressurized and the compressors do not run at all.

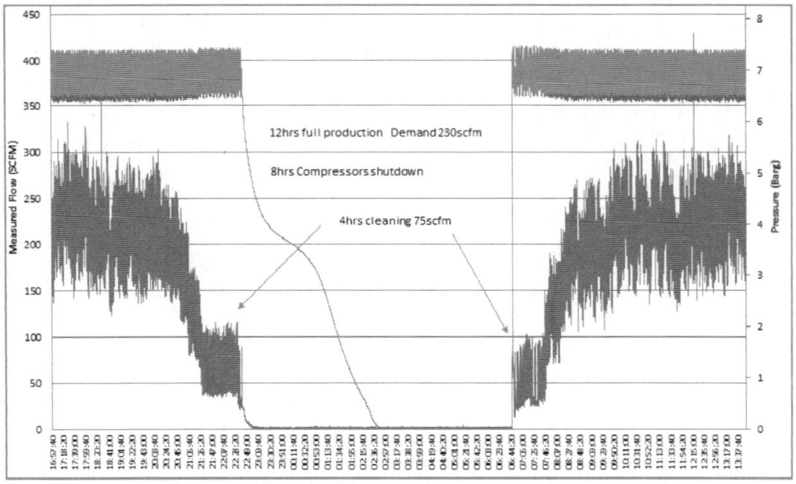

Figure 3.2. Air Flow and pressure characteristic.

(b) Pressure System

The compressor was set to load or unload on their internal pressure switches, there was no group controller. The individual settings are shown below:

	Unit	#1	#2	#3
Unload Pressure	*Barg*	*7*	-	-
Load Pressure	*Barg*	*6.8*	-	-

Figure 3.3 shows the characteristics of the pressure in the wet dry pressure under load and unload pressure. There was a pressure drop of 0.1 which was considered accepted.

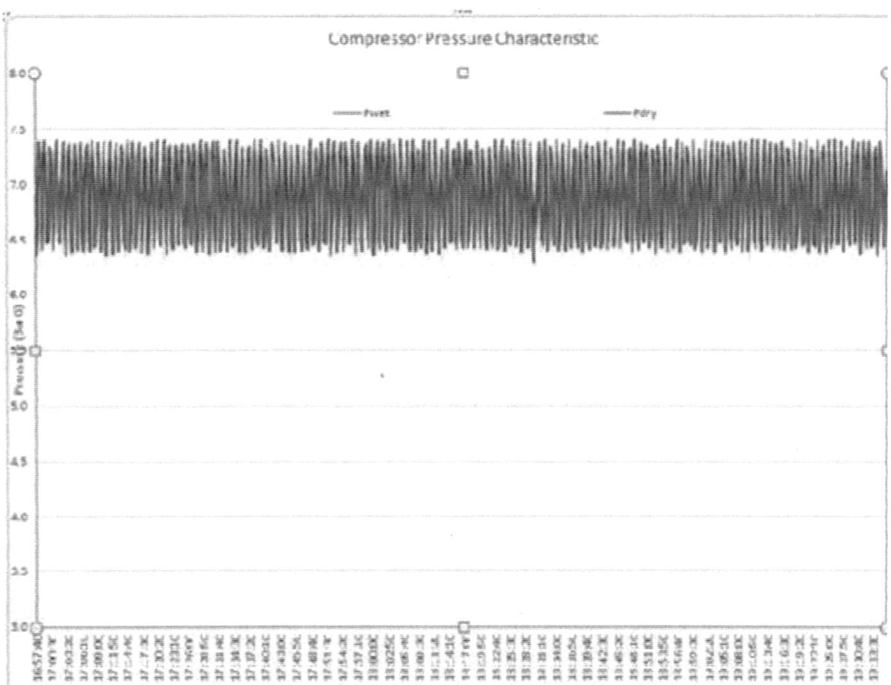

Figure 3.3. Compressor pressure characteristic.

A Curve performance of Pressure drop during full load and no load condition is shown in figure 3.4. We can see the individual pressure traces and the moving averages. The pressure drop is around *0.15barg* during production periods, dropping to around *0.5barg* when cleaning takes places due to the lessened flowing velocity through the treatment period. This level of pressure drop is acceptable within the best practice figure of *0.2barg* from wet to dry

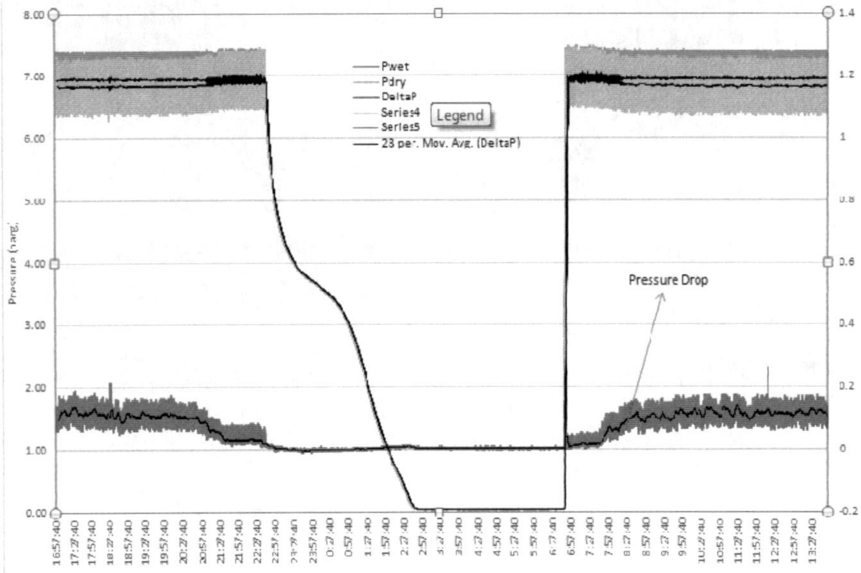

Figure 3.4. Compressor pressure drop.

(c) Air generation costs

Based on the measured data and conversation with the sample plant, the annual generation cost can be calculated.

Parameters used:

Unit Electricity Cost:	5p/kWh = 7.65c/kWh = IDR 1009.418/kWh
Annual Run Hours	-------(24/6/50) Production: 12hrs/6days/52weeks Cleaning: 4hrs/6days/52weeks
Mean Demand	-------scfm Production: 230scfm Cleaning: 75scfm

Figure 3.5 depicts the relationship between estimated generation costs with online capacity in cubic feet meter (*cfm*). There are differences the use of each compressor. The currently installed compressor, that operates GA-55, has the highest generation cost compare to others. It was found that the use of two GA-22 is more cost efficient.

IOP Conf. Series: Journal of Physics: Conf. Series **989** (2018) 012002 doi:10.1088/1742-6596/989/1/012002

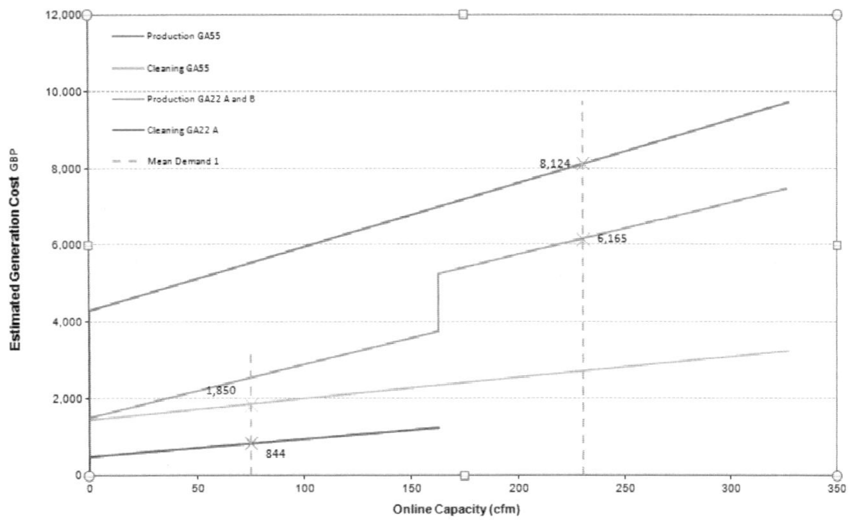

Figure 3.5.

(d) Leakage

A limited leak detection survey was carried out in the compressor and production house, using ultrasonic leak detectors. The total leak flow which can be saved is around *71cfm*, it is about 15.62 kW. Total Cost of Leakage was £3,899, it converted to dollars became $ 5,965.47 or 78,714,376.65 IDR. From the survey, the biggest leakage was from junction pipe at Compressor house which was *41cfm*, PI and UnCaser were *15cfm*. The action of repairing of this leak is a low-cost action which has big result in the reduction of energy consumption, so that the implementation needs to be taken seriously.

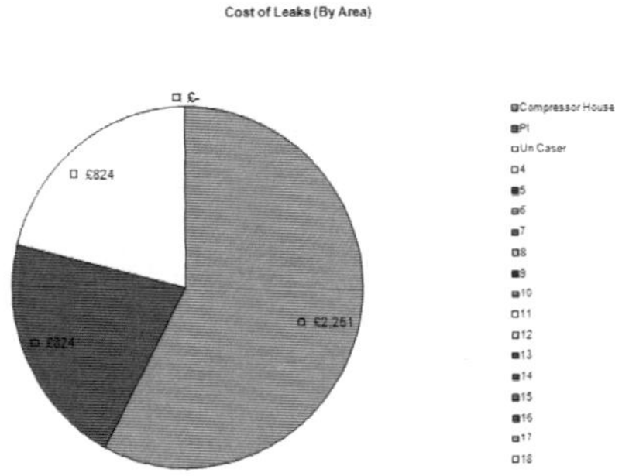

Figure 3.6a. Pie chart cost of leaks.

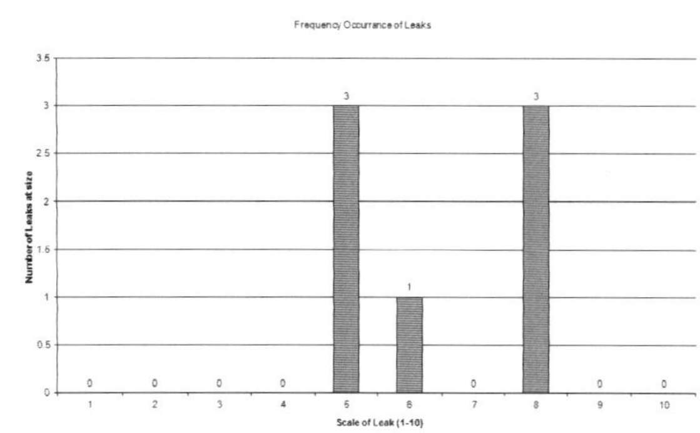

Figure 3.6b. Frequency occurence of leaks.

As we can see from figure 3.6a the blue chart refers to Compressor House, the purple one refers to PI site and cream colour one refers to UnCaser site. The biggest leakage among others was from Compressor House with the cost of leakage around £2,251 and both of PI and UnCaser site with the cost of leakage around £824. The frequency of occurrences of leaks is presented in figure 3.6b above. The number of occurrence of leaks frequency at Scale 5 and 8 are the same that is 3 times. These occurrences contribute big release air being produced.

No cost - low cost recommendations for improvements and savings

Some recommendation to no cost-low cost for improvement and saving:

- Detail calculation has been given to the cost saving was *78,714,377 IDR/yr.* with CO_2 savings was 46,200 Tonnes/yr. when site leakage is reduce. Reducing site leakage will reduce compressed air usage, reducing compressor power consumption is about *58,481 kWh/yr.* The locations which we focus to identify the leak are: Compressor House, PI, and Un Caser site. There are no risks to implementing leak reduction. For the next step, do site setup leak detection campaign.

- Repair two compressor type GA-22 and replace the Compressor GA-55 in term of applicable in use as production and cleaning. Cost of saving during a year was *141,500,145 IDR/yr.* with CO_2 being release was *110,737 Tonnes/yr.* and electrical power consumption was *140,174 kWh/yr.* The currently installed compressor, that operates GA-55, where it has the highest generation cost compare to others. The use of two GA-22 are more cost efficient compare with GA-55 and no Risk. Find the appropriate selection of compressor base on requirement process.

4. Conclusion

- It can be conclude that the total compressed air generating cost was related to total identified savings was 220,214,522 IDR/yr. with CO_2 savings was 156,937 Tonnes/yr. and power consumption was 198,655 kWh.
- No cost – Low cost action plan should be taken into account base on recommendation given with the 1st priority is leak reduction and 2nd is repair and running compressor GA-22 as soon as possible either for production or cleaning process.

5. References

[1] "UNIDO_and_energy_efficiency.pdf.".
[2] J. R. Neale and P. J. J. Kamp, "Compressed air system best practice programmes: What needs to change to secure long-term energy savings for New Zealand?," *Energy Policy*, vol. 37, no. 9, pp. 3400–3408, Sep. 2009.
[3] E. Consortium, *A Practical Guide to Energy Management of Facilities and Utilities*. Smithers Rapra Technology, 2011.

Acknowledgments

I wish to express my best gratitude to Maritje Hutapea as a Director of Renewable Energy in Directorate General of New Renewable Energy and Energy Conservation and Aris Ika Nugraha as a project coordinator UNIDO United Nations Industrial Development Organization in Indonesia for providing me an opportunity, guidance and encouragement to do my project research. I express my deepest thank as well to University of Udayana and IMEET research group members who give me a strong support to make this paper published. Thank you !

The Strategic Exercise of Options Using Government Subsidies: An Analysis of Production Subsidies for the Ground Source Heat Pump

Sheng-Hau Lin[1], Jia-Hsun Li[2*], Chih-Chen Hsu[3], Jing-Chzi Hsieh[4], Pin-Chao Liao [5]

[1] Ph D, Department of Ph. D Program in Civil and Hydraulic Engineering, College of Construction and Development, Feng Chia University, Taiwan, ROC.
[2] Assistant Research, School of Geographic and Oceanographic Sciences, Nanjing University, China.
[3] Chief Executive Officer, Professional Master's Program of Construction and Development, Feng Chia University.
[4] Professor, Department of Land Management, College of Construction and Development, Feng Chia University, Taiwan, ROC.
[5] Associate Professor, Department of Construction Management, Tsinghua University, China.
E-mail: candy01292001@gmail.com (Jia-Hsun Li); pinchao@tsinghua.edu.cn (Pin-Chao Liao)

Abstract. This study utilizes consolidation investment theory to incorporate with business strategies and government subsidy to develop a strategic exercise of options model. This empirical investigation examines the ground source heat pump (GSHP) government subsidy program, which is part of China's 12th Five Year Plan. The developed model is applied to explain the behaviours of business investment with regard to strategic investment timing, option values, and the influence of government subsidies in duopolistic real-world investment decisions. The results indicate that subsidy policy can reduce the differences of investment timing among GSHP investors and has clearly evidenced the positive benefit–cost ratio of government subsidy, which facilitates China's GSHP industry development.

1. Introduction

While the traditional real option approach (ROA) has been applied to analysing the optimal timing of execution, the lack of considerations to the interaction between two firms has been a remarkable deficiency. ROA ignores the potential impacts of other firms' exercise of strategies, which assumes that there are no impacts on the other firm's decision making, thus violating the reality of normal business strategic exercises. Therefore, this study develops an alternative real option games (ROG) model, which includes government subsidy within the framework of the strategic exercise of options to analyze the timing of competitors' strategic exercises.

The ROG analysis is an advanced investment decision technique that enables analysts to incorporate uncertain investment risks and strategic exercise of options by market participants. This study develops an ROG model by incorporating government subsidies to explain the behavior of strategic investment timings, option values, and interaction of market competitors with regard to the government subsidy policy in duopolistic real-world investment decisions. Dixit and Pindyck [1] reveal that a competitive environment may have a considerable impact on the valuation of real options, because when opponents exercise strategies that influence the value of the underlying asset, optimal policies cannot be determined in isolation as in classical models of real options. In particular, this environment, incorporating government subsidies, it possibly intervene the investment survivor, market participants' strategies, and their performance.

A subsidy is a form of financial or in-kind support extended to an economic sector generally with the aim of promoting economic and social policies. A government subsidy is an economic intervention that is commonly implemented to promote general welfare. Subsidies can be for various purposes, including subsidies for production, consumption, exports, employment, and taxes. This study focuses on real investment with production subsidies to ensure that producers are better off through either direct support or payments to factors of production. Fudenberg and Tirole [2] and Grenadier [3-6] highlight the importance of including strategic considerations in real option models as an advanced discipline of investment decisions. The ROG model was initially formulated based on a real option analysis (i.e., ROA) framework dependent on one or more stochastic underlying variables and from which the uncertain investment value can be obtained. The equilibrium of the investment game is employed to analyze the different market participants' strategic exercises of options to analyze their investment timings and option value. Aguerrevere [7] and Goto et al. [8] use ROG to analyze competition in the manufacturing sector. Angelou and Economides [9] use concept of ROG to analyze information and communication technology (ICT) business decision analysis model under the threat of competition. Hsu and Lambrecht [10] and Leung and Kwok [11] study the patent race within the framework of ROG, Huisman and Kort [12] and Aye and Fujiwara [13] employ ROG to analyze the timing of new technology adoption by all market participants, and Wu et al. [14] use a ROG evaluation method to explain competition in the thin film transistor liquid crystal display (TFT-LCD) industry.

ROG theory has been broadly employed to analyze real investment issues; however, to the best of our knowledge, there are no published studies that incorporate government subsidies in the analysis of market participants' investment timings and the option values. Central aspects unaddressed in the existing literature include under which conditions to conduct an investment, how much option values can be gained, and the influences of government subsidies on the investments. This study addresses the aforementioned questions within the framework of a continuous-time real options game approach.

This empirical investigation reviews the ground source heat pump (GSHP) government subsidy program, which is the main program of the 12th Five Year Plan (hereafter 12th FYP) with regards to accelerating the development of renewable technologies and encouraging their utilizations. The government of China plans to invest approximately US$12–US$18 billion to support the GSHP industry. The GSHP is a renewable energy technology that has increasingly attracted worldwide attention due to its advantages in energy efficiency and environmental friendliness [15-18]. The development of GHSP can help address the problem of air pollution and smog in China. The principle of GSHP means that renewable technology that uses energy stored in the ground to heat buildings and provide the hot water and element of cooling [19]. GSHP investment is an irreversible investment in an oligopolistic market due to its uneconomical cost-performance ratio compared with the traditional central heating, ventilation and air conditioning (HVAC) [20], and it generally receives government subsidies. Thus, it is an appropriate sample for this investigation.

The study is organized as follows. Section 2 provides an overview of the model including the assumptions and model formulations and determines the equilibrium set of exercise strategies for both market participants. Section 3 describes the empirical investigation: a case study of subsidies received in relation with GSHP in China. Section 4 presents the conclusion and policy implications.

2. Method: assumptions and model formulations

The model development of ROG includes the following assumptions: there are two identical investors in the market, each investor has an option to develop the same product, and the investment option can last forever. Hence, investors can choose to wait for optimal entry. The option values are affected by the strategies of other market participants. To address the relationship between the strategic exercise of options and government subsidies, we assume that the investor can receive a unit of government subsidy as long as a product can be sold. Upon the market reaching the Nash equilibrium, the investors' investment strategies are optimized and cross-interactive with each other. At the beginning of the investment game, each investor contemplates two choices: to be the pioneer to exercise the option (become the leader) or to postpone exercising the option (become the follower), with each strategies having an optimal window for "time to act."

The market price of a unit of product at time t, denoted by $P(t)$, equals the theoretical price $X(t)$, which is an exogenous shock process at time t, multiplied by the inverse demand function $D[Q(t)]$ fluctuates stochastically over time, plus the production government subsidy S, based on Equation (1):

$$P(t)=X(t)\times D\left[Q(t)\right]+S \tag{1}$$

where $D[Q(t)]$ is the inverse demand function and $Q(t)$ represents the states of industry supplies (market conditions). There are two market conditions, $Q(t)=1$ and $Q(t)=2$, which represent the market as a monopoly and duopoly, respectively. When the market is a duopoly, investors' strategies in the investment game are cross-interactive. Moreover, $D[Q(t)]$ also represents the deterministic contribution to the investment value function. We assume that the inverse demand function is downward sloping, $D[Q(t)]< 0$, to ensure "first-mover advantage" (i.e., FMA). Let $X(t)$ evolve following a geometric Brownian motion process:

$$dX(t)=\mu X(t)dt+\sigma X(t)dz, \tag{2}$$

where μ is the instantaneous conditional expected percentage change in $X(t)$ per time; σ is the instantaneous conditional standard deviation per unit time; and dz is the increment of a standard Wiener process for $X(t)$. Consequently, parameters μ and σ can be interpreted as the expected drift rate and volatility of log return, respectively. Upon the principle of risk neutrality, equation (2) can be rewritten as $dX(t)=(r-\delta)X(t)dt+\sigma X(t)dz$. δ is equal to risk premium, and r is risk-free return (i.e., the risk-free rate).

In the solution procedure, we first solve for the follower's optimal investment trigger and the option value function; then, we propose the analysis of the Nash equilibrium game.

2.1 Follower's trigger and option value

First, we model the follower's investment decision within the scenario of the leader's investment option, which has already been exercised; thus, the follower can construct an exercise strategy without considering the strategic exercise by market competitor(s). The follower's strategy is to maximize the option value as $F(X)$, which is equal to the follower's investment value. We found $F(X)$ by solving an equilibrium differential equation using a standard real option analysis technique and obtained an expression for the expected drift rate on the follower's option using Itô's Lemma. The following expression gives us the expected drift rate on the follower's option value $F(X)$ that solves differential Equation (3):

$$\frac{1}{2}\sigma^2 X(t)^2 F_{XX}+(r-\delta)X(t)F_X-rF=0 \tag{3}$$

when the consumer purchases a unit of a product, the market participants/investors receive a unit of S from the government. K is an exercise price which equal to the cost of investment. The differential Equation (3) must be solved subject to appropriate boundary conditions ensuring that the follower chooses the optimal trigger time (i.e., the "option to invest" is maximized):

$$\mathrm{F}(0)=0 \tag{4}$$

$$F(X_F^*)=X(t)D(2)+S-K \tag{4-1}$$

IOP Conf. Series: Journal of Physics: Conf. Series **989** (2018) 012003 doi:10.1088/1742-6596/989/1/012003

$$F_X(X_F^*) = D(2) \tag{4-2}$$

The first boundary condition $F(0) = 0$ means that zero is an absorbing barrier of exercise price. The second boundary condition is the "value-matching" condition, where the option value is $XD(2) + S - K$. The third boundary condition is the "smooth-pasting" or "high-contact" condition, which ensures that X_F^* triggers an entry strategy that maximizes the follower's option value; hence, X_F^* is regarded as the follower's optimal trigger timing (hereafter the follower's trigger). Through this procedure, we obtain X_F^* closed-form solutions for the value of $X(t)$, which triggers entry by standard dynamic programming methodology, and the follower's option value $F(X)$, in different steps:

$$X_F^* = \left[\frac{\beta}{\beta-1}\right] \times \left[\frac{K-S}{D(2)}\right], \tag{5}$$

$$F(X(t)) = \begin{cases} \left(\dfrac{K-S}{\beta-1}\right)\left(\dfrac{X(t)}{X_F^*}\right)^{\beta} & \text{for } X(t) < X_F^* \\ X(t)D(2) + S - K & \text{for } X(t) \geq X_F^* \end{cases} \tag{6}$$

The first row of (6) is the present option value when follower does not invest immediately. Its first term is the value of the option to invest. The latter term equals the NPV of the follower's investment discounted back from the (random) time of reaching the follower's triggers entry X_F^*. Consequently, $\left(X(t)/\ X_F^*\right)^{\beta}$ is interpreted as a stochastic discount factor, which is equal to the present value of $1 received when starting from level $X(t) < X_F^*$. The second row of (6) is the NPV of the profit flow when immediate investment is optimal [24].
where β is given by

$$\beta = \frac{1}{2} - \frac{(r-\delta)}{\sigma^2} + \sqrt{\left(\frac{(r-\delta)}{\sigma^2} - \frac{1}{2} + \frac{2r}{\sigma^2}\right.} > 1. \tag{7}$$

Given the above results, we express both the leader's and the follower's triggers and investment value functions in the pre-emption games, as illustrated in figure 1, by referring to Fudenberg and Tirole [2] and Azevedo and Paxson [25].

PROPOSITION 1. Conditional on the leader having begun investment in an investment game, the follower does not fear a threat of pre-emption and pre-gain from the government subsidy, which the leader already possesses. The follower's optimal strategy is to begin at $X(t)$. This is equal to or exceeds the trigger value X_F^*, where the investors' investment values are maximized. Figure 1 illustrates that the follower's investment is lower than the leader's investment below X_F^*.

2.2 *Leader's trigger and option value*

In the investment game, the leader retains pre-emption and is the first exerciser in the market. In a monopoly, the leader receives monopoly profits $X(t)D(1)$ [see Equation (8)] till the follower is triggered for entry. Hence, let $L(X(t))$ denote the leader's option value, which solves the equilibrium differential Equation (8):

$$0 = \frac{1}{2}\sigma^2 X(t)^2 L_{XX} + (r-\delta)\ X(t)L_X - rL + X(t)D(1)\,. \tag{8}$$

However, once the follower's option is exercised, the leader faces competition in the market and only receives duopoly profits. Differential Equation (8) must be solved subject to the boundary condition.

This boundary condition reflects the fact that the follower exercises the option at X_F^*, and the leader's option value equals that of the follower in the investment game :

$$L(X_F^*) = X_F^* D(2) + S - K \tag{9}$$

The solution for $L(X(t))$ can be expressed as

$$L(X(t)) = \begin{cases} X(t)D(1) + S - K + \left(\dfrac{\beta}{\beta-1}\right)(K-S)\left(\dfrac{D(2)-D(1)}{D(2)}\right)\left(\dfrac{X(t)}{X_F^*}\right)^{\beta} & \text{for } X(t) < X_F^* \\ X(t)D(2) + S - K & \text{for } X(t) \geq X_F^* \end{cases} \tag{10}$$

There exists a unique point X_L^* between 0 and X_F^* at which the option value for the leader and follower are equal. X_L^* is the point proposed by Fudenberg and Tirole [2] at which a duopoly market is in equilibrium. Further, Paxson and Pinto [26] indicate that till this point, either investor is better off being the follower; after this point, the leader is better off. Therefore, this point should be the leader's trigger X_L^*; however, a closed-form solution for X_L^* cannot be obtained because the resulting function is highly non-linear. Thus, the trigger can be obtained numerically by solving the following non-linear equation (where X_L^* is unknown):

$$X_L^* D(1) + S - K + \left(\frac{\beta}{\beta-1}\right)(K-S)\left(\frac{D(2)-D(1)}{D(2)}\right)\left(\frac{X_L^*}{X_F^*}\right)^{\beta} - \left(\frac{K-S}{\beta-1}\right)\left(\frac{X_L^*}{X_F^*}\right)^{\beta} = 0 \tag{11}$$

PROPOSITION 2. Conditional on the investors not yet beginning the investment game, when competition is added to the investment decision problem, other investors' actions are important factors that contribute to the leader's exercise strategy. Because of monopoly power, the leader tends to invest earlier to earn a government subsidy in advance. However, the leader may also be vulnerable to risk of loss in the early market stages. Figure 1 shows that the leader's option value is lower than that of the follower below X_L^* and higher than that of the follower between X_L^* and X_F^*. The leader's optimal strategy is to begin investment when $X(t)$ equals or exceeds the trigger X_L^*.

2.3 *Equilibrium exercise strategies*

Two types of equilibria can occur in the investors' choices of strategy—sequential and simultaneous equilibria—depending on the initial state of the theoretical price $X(t)$.

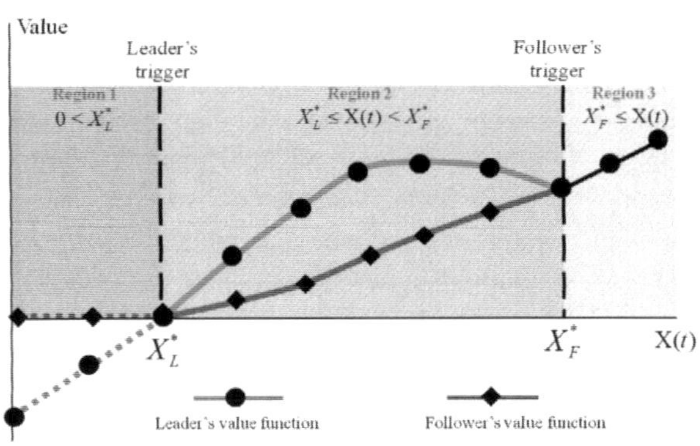

Figure 1. Investors' triggers and value in the pre-emption games
Source: Fudenberg and Tirole [2]; Azevedo and Paxson [25].

2.3.1 Equilibrium with sequential exercise: $X(t) < X_F^*$. Based on the *distribution* of $X(t)$ and the relative position of X_L^* and X_F^*, the price horizon can be divided into two different regions on the $X(t)$ line, $0 < X(t) < X_L^*$ and $X_L^* \le X(t) < X_F^*$, where sequential equilibrium occurs. These situations are described as follows:

Region 1: For $0 < X(t) < X_L^*$

If the investment game begins at the initial level, $X(t)$ is below X_L^*. As the market is not yet booming, the follower's option value is higher than the leader. (figure 1). Most investors prefer to be followers rather than leaders. If other investors have not yet entered, they initiate investment as soon as $X(t)$ equals or exceeds X_L^*. If their competitor has already entered, they wait till $X(t)$ rises to X_F^* before initiating an investment. Given the above strategy, sequential equilibrium will occur.

Region 2: For $X_L^* \le X(t) < X_F^*$

If the investment game starts from X_L^*, the benefits of a potentially temporary monopoly only equal the costs of early investment. The option value for both the leader and follower is equal at X_L^*, where each investor has a 50% chance of being the leader. If the other investor earns monopoly profits in the region where $X(t)$ is between X_L^* to X_F^*, they become the leader. As soon as one investor earns the leadership in the investment game, the follower chooses the optimal time at point X_F^* to invest.

If the investment game begins with an initial level of $X(t)$, which is between X_L^* and X_F^*, the market is more developed; the leader's option value is higher than that of the follower due to monopoly profits, and each investor would prefer to be the leader. If the leader's option has been exercised, the follower waits till the trigger X_F^* is reached.

2.3.2 Equilibrium with simultaneous exercise: $X_F^* \le X(t)$. We indicate that the investment game begins with an initial level of $X(t)$ and that each investor has the same investment value above X_F^*. In this range, any equilibrium will be characterized by a simultaneous exercise. If either investor's trigger $X(t)$ is equal to or greater than X_F^*, the other enters immediately thereafter.

3. Results: case study of the ground source heat pump subsidies in china

To demonstrate that this ROG model including government subsidy can be applied to real-world project evaluation under different scenarios, we investigate the GSHP subsidies specified in China's 12 FYP. The GSHP is a central heating and/or cooling system that transfers heat to/from the ground. It takes advantage of the moderate ground temperatures to boost efficiency and reduce the system's operational costs; moreover, it can be combined with solar heating to form a geo-solar system with even greater efficiency. The US Environmental Protection Agency (EPA) described the GSHP as the most energy-efficient, environmentally clean, and cost-effective space conditioning system available. Heat pumps offer significant emission reduction potential, particularly when used for both heating and cooling and when the electricity is produced from renewable resources. UK Environment Agency also indicated that ground pumps could help to achieve energy policy targets by 2020.

The data for the empirical investigation are sampled from *sunshi100.com* and describe a certain brand of GSHP central air conditioning. The observed sample unit services an indoor area up to 170 square meters. The parameters for empirical investigation are revealed in table 1.

We simulate a sensitivity analysis of investors' triggers by simulating $\mu, \sigma,$ *and* $X(t)$ and run simulations for the inverse demand function $D(1)$, the purchase price that consumers are willing to pay K, and government subsidies *S*. Importantly, the investors' cross-interactive strategic exercises and government subsidies *S* are also addressed in different scenarios and provide policy implications.

Table 1. Parameter definitions, base values, and simulations settings

	Parameters	Value and Simulation	Data Resources
$X(t)$	Theoretical price	Base value: - Simulations: US\$0–US\$100,000	-
K	Purchase price that consumers are willing to pay	Base value: US\$36,364 (CN¥224,000) Simulations: 0.5–5.0 times K	A certain brand of GSHP central air conditioning
$D(t)$	Inverse demand function	Base value: $D(1)=1$, $D(2)=0.75$ Simulations: $D(1)=$ 1–5	Refer to Grenadier (2000)
r	Risk-free rate	Base value: one-year government bonds yield (3.8% on September 1, 2014)	China Central Depository & Clearing Co., Ltd. (CCDC)
δ	Risk premium	Base value: 1.9%	Half of risk-free rate
μ	Expected drift rate	Base value: 1.9% Simulations: 0%–3.5%	Half of risk-free rate
σ	Volatility	Base value: 5% Simulations: 5%–35%	Lukas and Welling (2014)
S	Subsidies from government (People's Republic of China)	Base value: US\$10,909 (30% of purchase price that consumers willing to pay K) Simulations: 20%–50% times K	Subsidies policy of Beijing Municipal Commission of Development and Reform

Notes: This table lists the parameter definitions, base values, simulation settings, and data resources used to determine the leader's trigger X_L^*, the follower's trigger X_F^*, and the changes and differences between the two triggers.

3.1 *Sensitivity analysis of investors' triggers by simulating* μ

Figure 2 depicts the changes in X_L^* and X_F^* and the difference between them, G_δ, by simulating μ changes from 0%, 0.5%, 1.0%, 1.5%, 2.0%, 2.5%, and 3.0% to 3.5% under the scenario of r=3.8%, σ=5.0%, $D(1)$=1.0, $D(2)$=0.75, K=36,364, and S=10,909. The results of this simulation show that investors' triggers rise with increasing μ, regardless of the leader's or follower's triggers. G_μ also simultaneously increases (see figure 2), and with the increasing risk premium, the trigger rises sharply. When the risk premium change μ is 0%, the leader's and follower's triggers are US\$26,637 and US\$40,679, respectively, and the G_μ is US\$14,042. When the risk premium change μ is 0.5%, the leader's trigger is US\$28,270 and follower's trigger increase to US\$44,500 with a G_μ of US\$16,230. When μ is equal to 1.0%, the leader's and follower's reach US\$31,330 and US\$50,653, and the G_μ is US\$19,323. The leader's and follower's triggers are US\$36,516 and US\$60,294, respectively, with a G_μ of US\$23,778 when μ is 1.5%. When μ is equal to 2.0%, the leader's trigger increases to US\$45,246 and the follower's trigger rises to US\$75,913 with a G_μ of US\$30,667. When μ is equal to 2.5%, the leader's trigger reach US\$61,269 and follower's trigger increase to US\$104,052, respectively, and the G_μ is US\$42,783. When μ is equal to 3.0%, the leader's and follower's trigger increase to US\$97,916 and US\$69,945, respectively, with a G_μ of US\$69,945. In addition, the leader's and follower's triggers reach US\$257,777 and US\$445,212, respectively, with a G_μ of US\$187,435 when μ is 3.5%. The observed triggers and G_μ rise sharply, as shown in figure 2.

These results demonstrate that μ has a positive relationship with both the leader's and follower's triggers in the GSHP investment game. In other words, when the risk premium is high, potential investors are more likely to postpone exercising their investment options. The findings in this study are in line with previous studies on energy investment stimulus programs, such as Moon [23].

3.2 *Sensitivity* analysis of investors' triggers by simulating σ

Figure 3 show the changes in X_L^* and X_F^* and G_σ by simulating σ from 5%, 10%, 15%, 20%, 25%, and 30% to 35% under the scenario of r =3.8%, δ =1.9%, $D(1)$=1.0, $D(2)$ =0.75, K =36,364, and S =10,909. The simulation results indicate that when σ is 5%, the leader's trigger is US$43,095 and the follower's is US$72,098 with a G_σ of US$29,003. When σ is 10%, the leader's trigger increase to US$49,256 and follower's trigger increase to US$82,991. The G_σ is US$33,735. When σ is 15%, the leader's and follower's triggers increase to US$58,114 and US$98,534, respectively, with a G_σ of US$40,421. The leader's and follower's trigger are increase to US$69,275 and US$118,026, respectively, with a G_σ of US$48,751. When σ is 25%, the leader's trigger is US$82,663 and follower's is US$141,342 with a G_σ of US$58,678. When σ is 30%, the leader's and follower's triggers reaches US$98,303 and US$168,533, respectively, with a G_σ of US$70,230. When σ is 35%, the leader's and follower's triggers reaches US$116,243 and US$199,693, respectively, with a G_σ of US$83,450. This result demonstrates that uncertainty σ has a positive relationship with both the leader's and follower's triggers in the GSHP investment game.

The above property is consistent with the observations by Lukas and Welling [22] and Moon [23], who indicated that the higher the uncertainty (volatility), the higher the propensity to postpone exercising their investment options.

3.3 *Sensitivity analysis of investors' triggers by simulating* $X(t)$

Figure 4 shows the changes in $L(X)$ and $F(X)$ and the difference between them, $G_p = F(X) - L(X)$, by simulating the changes in $X(t)$ from 0 to 120,000, in step sizes of 10,000 and under the scenario of r =3.8%, δ =1.9%, σ =5%, $D(1)$ =1.0, $D(2)$ =0.75, K =36,364, and S =10,909. The simulation results can be divided into three regimes. In regime 1, $X(t) < X_L^* =$ US$43,095, When the theoretical price is US$10,000, the leader's and the follower's option values are US$−15,886 and US$685, respectively, with a G_p of is US$16,571. The leader's and the follower's option values are US$−7,053 and US$2,538, respectively, and G_p is US$9,591 when the theoretical price is US$20,000. When the theoretical price is US$30,000, the leader's and the follower's option values are US$1,107 and US$5,459, respectively, with a G_p of is US$4,352. As the market is not yet booming, the leader's option value is lower than that of the follower due to weak initial market conditions. Hence, the leader faces an investment loss. Most investors prefer to be followers rather than leaders. However, with increasing theoretical price $X(t)$, when the theoretical price reaches X_L^*, all investors' option values equal US$10,824, with a G_p of is zero; then, investors in the GSHP market decide to invest immediately.

When the theoretical price is located between X_L^* and X_F^*, firms follow regime 2: the leader's and the follower's option values are US$15,518 and US$14,332, respectively, and G_p is US$−1,186 when the theoretical price is US$50,000. When theoretical price reaches US$60,000, the leader's and the follower's option values are US$21,806 and US$20,227, respectively, with a G_p of US$−1,579. When the theoretical price is increased to US$70,000, the leader's option value increases to US$27,499 and the follower's option value increases to US$27,066. The G_p reduces to US$−433. In this regime, the investment market is more developed, the leader enjoys the profits from pre-emption;

thus, the leader's option value is higher than that of the follower, and each investor would prefer to be the leader.

When $X(t)$ is higher than X_F^*= US$72,098, firms follow regime 3: option values of both investors are equal, their option values increase to US$28,619, and G_p is zero. When theoretical price reach US$80,000, the investor's option values increase to US$34,545, with a G_p of US$0. When theoretical price increase to US$90,000, the leader's option value increases to US$42,045, follower's option value also increase to US$42,045, with a G_p of US$0. The investors' option values are equal, their option values increase to US$49,545, and G_p is US$0 when the theoretical price is US$100,000. When theoretical price is increased to US$110,000, the all investors' option values are both US$57,045, the difference is zero. The investors' option values are US$64,545 when theoretical price increase to US$120,000. In this range, each investor would prefer to enter immediately. If either investor's trigger is equal to or greater than X_F^*, the other also enters immediately thereafter.

The empirical investigation demonstrates that the developed model is consistent with those in existing studies on standard real option games, such as Fudenberg and Tirole [2], Pawlina and Kort [24], and Azevedo and Paxson [25]. Thus, this empirical investigation supports the use of the developed model in the practical analysis of GSHP investments.

3.4 Sensitivity analysis of investors' triggers by simulating $D(1)$

Figure 5 show the changes in X_L^* and X_F^*, and the difference, $G_{D(1)}$, by simulating the changes in $D(1)$ from 1.0, 1.5, 2.0, 2.5, 3.0, 3.5, 4.0, and 4.5 to 5.0 under the scenario of r=3.8%, δ=1.9%, σ=5%, $D(2)$=0.75, K=36,364, and S=10,909. First, we set the follower's trigger X_F^* at US$72,098. The simulation result shows that the leader's trigger is US$43,095 when the monopoly inverse demand function $D(1)$ is 1.0, and $G_{D(1)}$ is US$29,003. When $D(1)$ is 1.5, the leader's trigger decrease to US$23,714. The $G_{D(1)}$ is US$48,384. When $D(1)$ is 2.0, the leader's trigger decrease to US$16,307, and the difference increases to US$55,791. The leader's trigger fall to US$12,410, and $G_{D(1)}$ is US$59,688. When $D(1)$ is 3.0, the leader's triggers decrease to US$10,010, and difference increases to is US$62,088. When $D(1)$ is 3.5, the leader's trigger decrease to US$8,385, and $G_{D(1)}$ increases to is US$63,713. When $D(1)$ is 4.0, the leader's trigger decrease to US$7,212, and difference increases to is US$64,886. When $D(1)$ is 4.5, the leader's trigger decrease to US$6,326, and $G_{D(1)}$ increases to is US$65,772. When $D(1)$ is 5.0, the leader's trigger decreases to US$5,633 and the $G_{D(1)}$ increases to US$66,465. This result means that a higher $D(1)$ reduces the leader's trigger due to strong demand and government subsidy in a monopoly market. In other words, the stronger inverse demand function $D(1)$ promotes investors to be the leader. When considering the traditional real option, investors select an entry trigger that maximizes the investment value because the inverse demand function is always identical. In this study, we add the monopoly inverse demand function to determine the impacts of investors' option value, triggers, and strategies. We obtain more investment information using a model of the real option game compared with traditional real options.

3.5 Sensitivity analysis of investors' triggers by simulating K

Figure 6 show the changes in X_L^* and X_F^* and the difference between them, $G_K = X_F^* - X_L^*$, by simulating the changes in K from 0.5, 1.0, 1.5, 2.0, 2.5, 3.0, 3.5, 4.0, and 4.5 to 5.0 times purchase price that consumers are willing to pay (exercise price) under the scenario of r=3.8%, δ=1.9%, σ=5%, $D(1)$=1.0, $D(2)$=0.75, and S=10,909. The results of this simulation reveal that the leader's and follower's triggers are US$21,548 and US$36,049, respectively, with a G_K of US$14,052 when consumers are willing to pay 0.5 times the purchase price. When K increase to 1.0 times the purchase price, the leader's and follower's triggers are US$43,095 and US$72,098, respectively, with a G_K of US$29,003. Investor's triggers and G_K are gradually increased along with increasing purchase price K. When K increases to 1.5 times the purchase price, the leader's and follower's triggers are US$64,643 and US$108,148, respectively, and G_K rises to US$43,505. When K increase to 2.0 times the purchase price, the leader's and follower's triggers are US$86,191 and US$144,197, respectively, with a G_K of US$58,006. When K increase to 2.5 times the average purchase price, the leader's and follower's triggers are US$107,738 and US $180,246, respectively, with a G_K of US$72,508. In addition, the leader's and the follower's trigger are US$129,286 and US$216,295, respectively, with a G_K of US$87,009 when the purchase price that consumers are willing to pay is 3.0 times the purchase price. When purchase price is 3.5 times the purchase price, the leader's and the follower's trigger are US$150,833 and US$252,344, respectively, with a G_K of US$101,511.The leader's and the follower's trigger are US$172,381 and US$288,393, respectively, and G_K is US$116,012 when the purchase price that consumers are willing to pay is 4.0 times the average purchase price. When K reaches to 4.5 times the purchase price, the leader's and follower's triggers are US$193,929 and $324,443, respectively, with a G_K of US$130,514. When K increases to 5.0 times the purchase price, the leader's and follower's triggers are US$215,476 and $360,492, respectively, and G_K increases to US$145,015. Depending on the purchase price that consumers are willing to pay, K, higher costs (exercise price) postpone investment decisions, which is consistent with the results in Moon [23] that indicated investors wait longer if the costs are increased.

3.6 Sensitivity analysis of investors' triggers by simulating s

Figure 7 shows the changes in X_L^* and X_F^*, and $G_S = X_F^* - X_L^*$, by simulating the changes in s from 0%, 5%, 10%, 15%, 20%, 25%, 30%, 35%, 40%, and 45% to 50% times the purchase price that consumers are willing to pay, under the scenario of r=3.8%, δ=1.9%, σ=5%, $D(1)$=1.0, $D(2)$=0.75, K=36,364, and s=10,909. The simulation result reveals that the leader's trigger is US$61,565, the follower's is US$102,998, and G_S is US$41,433 within the scenario of zero government subsidies. When government subsidies rise to 5% times the purchase price that consumers are willing to pay, the leader's and follower's triggers decrease to US$58,486 and US$97,848, respectively, and G_S decreases to US$39,361. The leader's and follower's triggers decrease to US$55,408 and US$92,698, respectively, with a G_S of US$37,290 when the government subsidies reach to 10% times the purchase price that consumers are willing to pay. When government subsidies increase to 15% times the purchase price, the leader's and follower's triggers decrease to US$52,330 and US$87,548, respectively, and G_S decreases to US$35,218. The leader's and follower's triggers decrease to US$49,252 and US$82,398, respectively, with a G_S of US$33,146 when the government subsidies reach to 20% times the purchase price that consumers are willing to pay. When government subsidies rise to 25% times the purchase price, the leader's and follower's triggers decrease to US$46,174 and US$77,248, respectively, and G_S decreases to US$31,075. In GSHP market, investor's triggers and G_S are gradually decreased along with increasing government subsidies. When government subsidies reach to 30% times the purchase price, the leader's and follower's triggers decrease to US$43,095 and US$72,098, respectively, with a G_S of US$29,003. When government subsidies rise to 35% times the average sample price, the leader's and follower's triggers decrease to US$40,017 and US$66,948, respectively, and G_S decreases to US$26,931. When government subsidies rise to 40% times the purchase price, the leader's and follower's triggers decrease to US$36,939 and US$61,799, respectively, and G_S decreases to US$24,860. When government subsidies rise to 45% times the purchase price, the leader's and follower's triggers decrease to US$33,861 and US$56,649, respectively, and G_S decreases to US$22,788. When the government subsidies rise to 50% times the purchase price, the leader's and follower's triggers fall to US$30,782 and US$51,499, respectively, and G_S decreases to US$20,716. The above simulation demonstrates the negative relationship between investors' triggers and government subsidies. In other words, higher government subsidies attract GSHP investors to enter the market earlier. In other words, more investors are attracted to and more likely to invest because of government subsidies. Investors stimulate the market as they are more willing to join the GSHP market and produce more GSHP products, thus achieving the goal of government policy.

3.7 The effects of GSHP subsidy policy on investors' option value and government's benefit–cost analysis (benefit–cost analysis, BC ratio)

To examine China's subsidy policy (30% times the average purchase price that consumers are willing to pay) and how to influence the investors' option value and its benefit and cost for the goal of promoting the GSHP industry, we simulate the investors' option values under the scenario of S = 28%-32% k. Therein, we display changes in the leader's and follower's option values under the scenario of S = 28%—32%, $L(X)$, $F(X)$, as well as the difference in option values, G_L, and G_F, and the benefit ratio, B from adding government subsidies of 1% by simulating the theoretical price changes in $X(t)$ from 0 to 90,000, in step sizes of 10,000 and under the scenario of r =3.8%, δ =1.9%, σ =5%, $D(1)$=1.0, $D(2)$=0.75, and K =36,364.

Under the scenario of S = 28% k, the leader's and the follower's option values are US$−7,741 and US$2,475, respectively, when the theoretical price is US$20,000, and when government adds subsidies of 1% to –29% k, the investors' option values increases to US$−7,397 and US$2,506, at which point the differences in option values are US$344 and US$31, respectively. When the theoretical price reaches US$60,000, the differences in option values are US$208 and US$247, respectively. Evidently, the leader's (follower's) option value experiences a decrease (increase) when adding the theoretical price under the same scenario of subsidy policy. Hence, the results conclude with certainty that the policy of government subsidy can decrease the differences in option values between investors of GSHP.

As previously mentioned previously, we will display the changes in the benefit–cost ratio under the different scenarios in examining the effect of China's subsidy policy on GSHP investors' option values and its implication of benefit and cost for subsidy policy. When the theoretical price of GSHP is US36,364 (i.e. the average purchase price that consumers are willing to pay), the leader's and follower's values are US$5,662 and US$7,754, respectively and differences of option values are US$303 and US$96, respectively, under the scenario of a 29% subsidy. The government can earn 1.098 times the benefits for promoting overall GSHP industry when it increases costs by 1% (i.e. subsidy of 1%). Under the current China subsidy policy (i.e. 30%) and the same level of the theoretical prices of GSHP, the benefit–cost ratio increases to 1.100, even though the difference in subsidy is also equal to 1%. The results indicate a positive and strong relationship between subsidy policy and promoting the GSHP industry. More specifically, China's current GSHP production subsidies policy will earn 1.1 times the benefits for GSHP industry development.

Moreover, we also discover that the benefit–cost ratio will exhibit an increasing trend with "increasing theoretical price of GSHP with same subsidy policy" or "increasing subsidy policy with same theoretical price of GSHP". For the first case, when the theoretical price rises to US$20,000 from US$10,000, the benefit–cost ratio increases to 1.032 from 1.009 (1.032−1.009 = 0.023). When the theoretical price rises to US$70,000 from US$60,000, the benefit–cost ratio increases to 1.336 from 1.251 (1.336−1.251 = 0.085). The results suggest that subsidy policy may have a stronger impact on the higher GSHP price. For the second case, when the theoretical price rises to US$70,000 from US$60,000, the benefit–cost ratio increases to 1.336 from 1.251 (1.336−1.251 = 0.085), and 1.365 from 1.273 (1.365−1.273 = 0.092) under the scenarios of subsidy ratios of 28% and 32%, respectively. These findings support the claim that higher subsidy policy will create more benefits for industry development. Thus, the government is able to expand the economic sector through subsidies to promote the GSHP and energy industries in China.

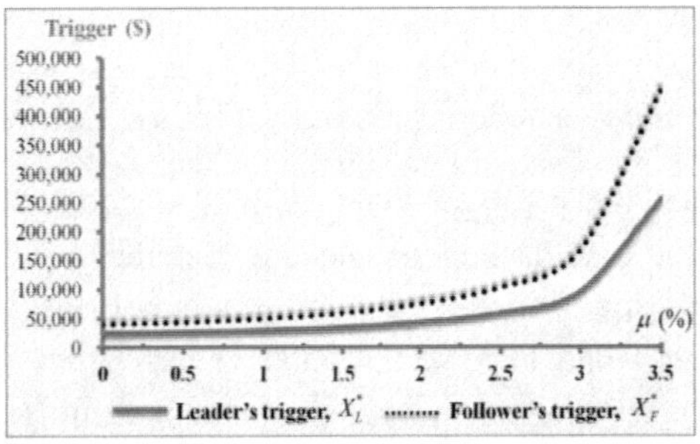

Figure 2. Sensitivity analysis of investors' triggers by simulating expected drift rate μ

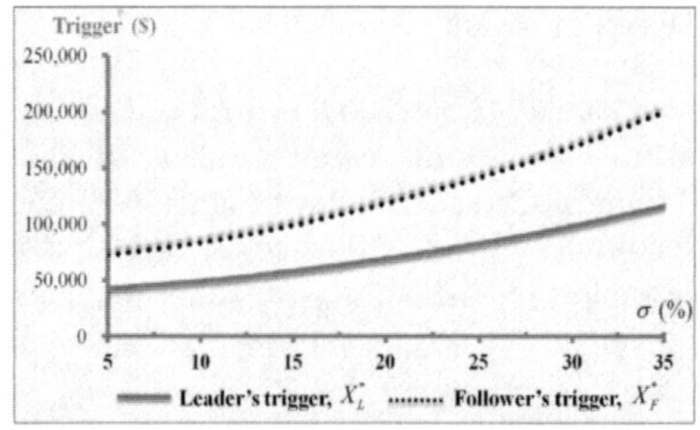

Figure 3. Sensitivity analysis of investors' triggers by simulating volatility σ

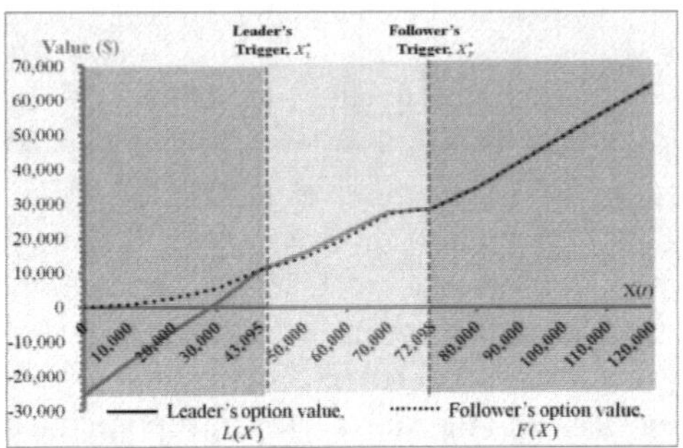

Figure 4. Sensitivity analysis of investors' option value by simulating theoretical price $X(t)$

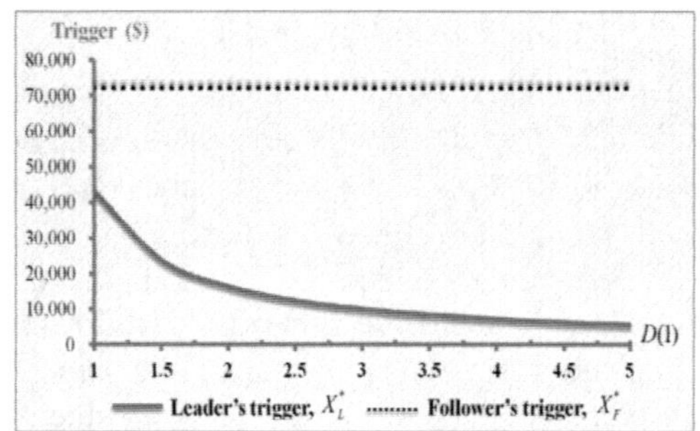

Figure 5. Sensitivity analysis of investors' triggers by simulating inverse demand function $D(1)$

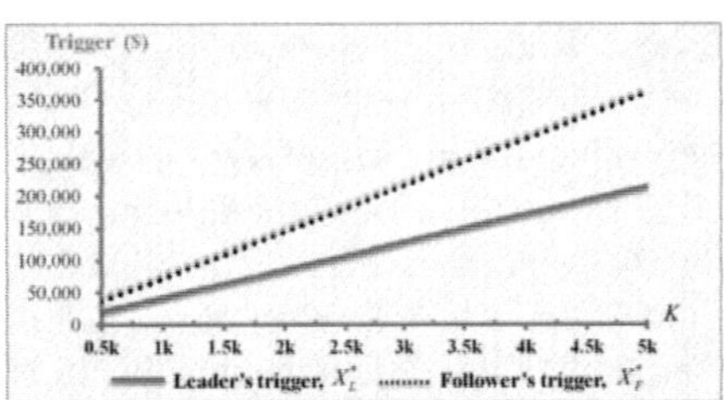

Figure 6. Sensitivity analysis of investors' triggers by simulating purchase price that consumers are willing to pay K

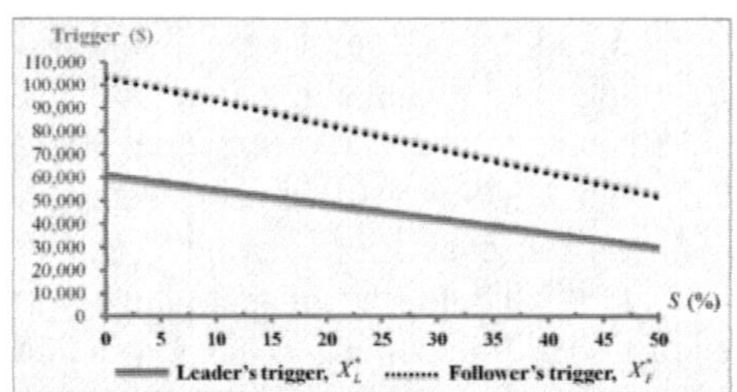

Figure 7. Sensitivity analysis of investors' triggers by subsidies S

Table 2. The effects of GSHP subsidy policy on investors' option value and government's benefit–cost analysis

	$S = 30\% K$		Benefit Cost ratio
$X(t)$	$L_{30\%}(X)$ $(G_{L2} = L_{30\%}(X) - L_{29\%}(X))$ (in US$)	$F_{30\%}(X)$ $(G_{F2} = F_{30\%}(X) - F_{29\%}(X))$ (in US$)	$B_{30\%-29\%} = \frac{(G_{L2}+G_{F2})}{1\%K}$
0	-25,455 (364)	0 (0)	1.000
10,000	-15,886 (358)	685 (9)	1.009
20,000	-7,053 (344)	2,538 (32)	1.032
36,364	**5,964 (302)**	**7,853 (98)**	**1.100**
40,000	8,624 (289)	9,402 (118)	1.120
50,000	15,518 (250)	14,332 (180)	1.183

	$S = 30\% K$		Benefit Cost ratio
$X(t)$	$L_{30\%}(X)$ ($G_{L2} = L_{30\%}(X) - L_{29\%}(X)$) (in US$)	$F_{30\%}(X)$ ($G_{F2} = F_{30\%}(X) - F_{29\%}(X)$) (in US$)	$B_{30\%-29\%} = \frac{(G_{L2} + G_{F2})}{1\% K}$
60,000	21,806 (204)	20,227 (254)	1.258
70,000	27,499 (150)	27,066 (339)	1.345
80,000	34,545 (364)	34,545 (364)	2.000

Notes: To examine China's subsidy policy (30% times the average purchase price that consumers are willing to pay) and how to influence the investors' investment value and its benefit and cost for the goal of promoting GSHP industry, we depicts the changes in leader's and follower's option values under the scenario of S = 30%, $L(X)$, $F(X)$ as well as the difference in option values, G_L and G_F, and benefit ratio, B from adding government subsidies of 1%, by simulating the theoretical price changes in $X(t)$ from 0 to 80,000 in step sizes of 10,000 and under the scenario of r =3.8%, δ =1.9%, σ =5%, $D(1)$=1.0, $D(2)$=0.75, and K=36,364.

4. Discussion

The above results show that investors time their investments according to increasing expected drift rate and volatility of market participants (leaders or followers). The aforementioned results are consistent with those of existing studies [22, 23]. In addition, a higher inverse demand function increases the leader's option value and enhances investors' desires to trigger earlier GSHP investment due to its monopolistic power. The higher purchase price that consumers are willing to pay (exercise price) also postpone investors' decision timing, which is consistent with the results in Moon [23].

With respect to the discussion of government subsidies, the aforementioned results are on the whole the consistent with those of existing studies [20, 21], but it is not entirely identical. Kim et al. [21] propose a framework for determining government subsidies for the installation of alternative energy systems using real options from the viewpoint of households and indicates that a proper level of government subsidy will effectively encourage households to install such energy systems. Huang et al. [20] investigate developer and consumer payoffs and strategies in GSHP development using the evolutionary game model. Their findings reveal that subsidies are the most efficient manner for promoting investment in GSHP systems. The present study enhances the previous findings by including government subsidies for energy system installations from the investor's viewpoint unlike Kim et al. [21]. We also obtain more information regarding investors' triggers, option values, and strategies using a real option game model compared with the traditional real option method or game theory model. This study's results, similar to those of Huang et al. [20], imply that higher government subsidies enhance investors' desires and trigger GSHP investment. Significantly, the evidence indicates that more valuable subsidies promote investors to make earlier investments and reduce the difference between them; thus, the differences decrease with increasing government subsidies, investors are more likely to invest, and more investors are attracted to boost the investment because the government subsidies encourage more GSHP products in the market and achieve the goal of government policies to promote the energy industry. Examining currently China's GSHP production subsidies policy from viewpoint of Benefit Cost ratio, this study also discovers that government can earn 1.1 times benefits for GSHP industry development (see table 2).

5. Conclusion and policy implications

This proposed model is the first attempt to incorporate the real option game with subsidies into investors' decision making for the development of new products from the investors' viewpoint. Utilizing consolidation investment theory, this study considers investment-related business strategies and develops a relevant strategic option model. The model is applied to explain the behaviors of energy industries with regard to strategic investment timing, option values, and the influence of government subsidies in duopolistic real-world investment decisions. This empirical investigation examined the GSHP government subsidy program, which is part of China's Twelfth Guideline (2011). The GSHP is part of the green energy industries, which often utilize government subsidies because they are unable to maintain economic viability. The empirical investigation simulates the influences of interior parameters, $\mu,\ \sigma,\ and\ X(t)$, and of underlying and exterior parameters, $D(\bullet),\ K,\ and\ S$. The results show that investors' triggers and the difference between these triggers rise with the increasing risk premium and volatility of market participants (leaders or followers). Concerning the simulative influence of $X(t)$, when $X(t) < X_L^*$, due to weak initial market conditions, the leader faces a loss due to weak initial market conditions. Further, when $X_L^* \leq X(t) < X_F^*$, the leader enjoys the profits arising from pre-emption and has an option value higher than that of the follower. In addition, when $X_F^* \leq X(t)$, both investors' option values are equal. The investigation demonstrates the consistency of the developed model with those in relevant existing studies, such as Fudenberg and Tirole [2], Pawlina and Kort [24], and Azevedo and Paxson [25].

Moreover, we find that a higher inverse demand function also increases the leader's triggers due to monopolistic power. Further, a higher purchase price postpones the investment timing. The government subsidy enhances investors' desires to trigger earlier GSHP investment. The 12th FYP includes targets for developing energy efficient technologies; increasing the share of renewable forms of energy; and reducing emissions, carbon intensities, and pollution [27]. China's government invests approximately US$12–US$18 billion in the GHSP government subsidy program, which is highly emphasized in the 12th FYP. This study demonstrates that the more valuable a government subsidy the more it diminishes the differences of triggers (investment timing) or option values between GSHP investors. This empirical study also clearly evidenced that China's current GSHP production subsidies policy will earn 1.1 times the benefits for the GSHP industry, and hence can effectively enhance GSHP industry development, which also means that the government subsidy is effectively achieving its goal for increasing the share of renewable forms of energy and is able to expand the economic sector through subsidies to promote the GSHP industry. The aforementioned results are on the whole consistent with those of existing studies [20, 21].

6. References

[1] Dixit AK, Pindyck RS. *Investment Under Uncertainty*. Princeton University Press; 1994.

[2] Fudenberg D, Tirole J. Pre-emption and rent equalization in the adoption of new technology. *Rev Econ Stud* 1985; 52(3): 383–401.

[3] Grenadier SR. The strategic exercise of options: development cascades and overbuilding in real estate markets. *J Financ* 1996; 51(5): 1653–1679.

[4] Grenadier SR. Information revelation through option exercise. *Rev Financ Stud* 1999; 12(1): 95–129.

[5] Grenadier SR. Option exercise games: the intersection of real option and game theory. *J Appl Corp Financ* 2000; 13(2): 99–107.

[6] Grenadier SR. Option exercise games: an application to the equilibrium investment strategies of firms. *Rev Financ Stud* 2002; 15(3): 691–721.

[7] Aguerrevere F. Equilibrium investment strategies and output price behavior: a real options approach. *Rev Financ Stud* 2003; 16(4): 1239–1272.

[8] Goto M, Takashima R, Tsujimura M, Ohno T. *Entry and exit decisions under uncertainty in a symmetric duopoly*. Presented at the real options conference 2008. Rio de Janeiro; 2008.

[9] Angelou GN, Economides AA. A multi-criteria game theory and real-options model for irreversible ICT investment decisions. *Telecommun Policy* 2008; 10–11: 686–705.

[10] Hsu Y, Lambrecht B. Pre-emptive patenting under uncertainty and asymmetric information. *Ann Oper Res* 2007; 151: 5–28.

[11] Leung C, Kwok Y. Patent-investment games under asymmetric information. *Eur J Oper Res* 2012; 223: 441–451.

[12] Huisman K, Kort P. Strategic technology adoption taking into account future technological improvements: a real options approach. *Eur J Oper Res* 2004; 159: 705–728.

[13] Aye NN, Fujiwara T. Application of option-games approach to the irreversible investment for a new energy industry in Myanmar by simple one-stage strategic model: focused on potential of smart house. *Glob J Flex Syst Manag* 2014; 15(3): 191–202.
[14] Wu LC, Li SH, Ong CS, Pan C. Options in technology investment games: the real world TFT-LCD industry case. *Technol Forecast Soc* 2012; 79: 1241–1253.
[15] Yang WY, Zhou J, Xu W, Zhang G. Current status of ground-source heat pumps in China. *Energy Policy* 2010; 38: 323–332.
[16] Yang Z, Heng C, Xi W, Yihuang C. Research on improving energy efficiency and the annual distributing structure in electricity and gas consumption by extending use of GEHP. *Energy Policy* 2011; 39: 5192–5202.
[17] Yuan Y, Cao X, Sun L, Yu N. Ground source heat pump system: a review of mimulation in China. *Renew. Sust Energy Rev* 2012; 16: 6814–6822.
[18] Yang W. Experimental performance analysis of a direct-expansion ground source heat pump in Xiangtan, China. *Energy* 2013; 59: 334–339.
[19] Omer AM. Ground-Source Heat Pumps Systems and Applications. *Renew Sust Energy Rev* 2008; 12: 344–371.
[20] Huang Y, Liao PC, Tsai CH, Gui S. Modeling the relationships of factors affecting dissemination of ground source heat pump (GSHP) in China. *Adv Mater Res* 2013; 723: 976–984.
[21] Kim B, Lim H, Kim H, Hong, T. Determining the value of governmental subsidies for the installation of clean energy systems using real options. *J Constr Eng Manag* 2012; 138: 422–430.
[22] Lukas E, Welling, A. On the investment-uncertainty relationship: a game theoretic real option approach. *Financ Res Lett* 2014; 11(1): 25–35.
[23] Moon Y. Optimal time to invest energy storage system under uncertainty conditions. *Energies* 2014; 7: 2701–2719.
[24] Pawlina G, Kort PM. Real option in an asymmetric duopoly: who benefits from your competitive disadvantage? *J Econ Manag Strat* 2006; 15(1): 1–35.
[25] Azevedo A, Paxson D. Developing real option game models. *Eur J Oper Res* 2014; 237(3): 902–921.
[26] Paxson D, Pinto H. Rivalry under Price and Quantity Uncertainty. *Rev Financ Econ* 2005; 14: 209–224.
[27] Thomson E, Introduction to special issue: Energy issues in China's 12th Five Year Plan and beyond. *Energy Policy* 2014; 73: 1–3.

Acknowledgements
The authors thank the National Natural Science Foundation of China (No. 51578317) for their support of this study. The authors are also grateful for input from industry professionals who participated in this research.

Thar Coalfield: Sustainable Development and an Open Sesame to the Energy Security of Pakistan

Adven Masih[1]

[1] PhD candidate Ural Federal University, Ekaterinburg, Russia
E-mail: adven.masikh@urfu.ru

Abstract. The paper discusses the role of Thar-coalfield, a 175 Billion tones reserve in enhancing the energy and combating global environmental change from the local and regional aspects. Pakistan's energy requirements are potentially huge. Being the sixth largest country in the world, with its growing population exceeded 190m by 2015. Rising population, improved living standards, increased per capita energy use, and industrialization has led to a high energy demand growth. According to latest reports the gap between the demand and supply of electricity is around 6,000MW. To meet the projected demand exploiting indigenous resources, such as Thar coalfield, a 100,000MW generation capacity reserve, could be the possible answer. Due to sustainable techniques in energy sector, 1) Coal mining is moving towards sustainable development; 2) circular economy has proven useful concept for promoting sustainable development; 3) coal industry can minimize its environmental impact from local to global level.
Besides energy goals, environmental degradation associated with the mining activity poses a serious threat to the region. Therefore, some challenges need to be addressed, e.g., discharge management issues, concerns regarding pollution control, lack of technology needed to replenish solid waste; and, increased socioeconomic and environmental pressure on the coal industry. The study discusses how sustainable development measures in Thar coalfield can run the engines of economic growth without hurting the natural environment promoting prosperity in Pakistan.

1. Introduction

For decades, the mining industry has remained a hot arena worldwide for academic and public policy research in terms of sustainable development [1, 2]. Mining industry has played a significant role in economic growth and poverty alleviation by providing millions of jobs to the locals [3, 4].
According to World Bank, Pakistan's per capita energy supplied level, the per-capita total primary energy supply (TPES) in the country was as low as half a ton of oil equivalent (toe) in 2014, which is about over a quarter of the global average, and about one-fifth that of China makes it clear that the country has to increase its energy resources against the improved living standards of its citizens. Electric-power supply has a close connection with modern energy services (such as lighting, refrigeration, and space cooling) and other services such as water treatment, communications and many aspects of transportation. The per capita availability of electric power in Pakistan was almost 472 kWh in 2014, which is almost half of world's average and just over one-seventh of China [5].
Pakistan has long suffered from an insufficient supply of electricity in relation to the demand in 2013-2014, the total shortage of energy was estimated to be 4000MW i.e.19%, while during peak hours the

shortage was as high as 7000MW which is 32% of total demand for electricity [6]. The quality of power supply in the country is also very poor, with unstable voltages and routine frequency excursions. In fact, the inadequate and unreliable supply of electricity has often considered as the main obstacle to industrial development.

Over 140 million Pakistanis suffer a 12 hours of daily power outages or they don't have an access to the power grid and the cost of these no electricity hours reaches $2.3bn annually spent in the form of candles and kerosene oil [6]. Pakistan's exceptional and stable growth of 6% GDP per annum from 2002 to 2007 increased the demand of electricity. However due to improper energy sector planning the country faced severe power cuts resulted in 2.5% GDP loss, unemployment of 0.535 million workers and a loss of $1.3 billion in exports of the country [7].

Pakistan's electricity supply is provided by WAPDA, IPPs (Independent Power Producers), GENCOs, KESC and PAEC (Pakistan Atomic Energy Commission). According to National Transmission and Dispatch Company 2015 report, IPPs (Thermal) contribute up to 9085 MW which mainly rely on gas and oil power plants. WAPDA the hydroelectric based power plants have a capacity of 6902 MW, whereas GENCOs shares 7663 MW to national grid. Apart from these major sources of electricity, other sources such as wind, Nuclear (PAEC) and coal also contribute few hundreds of megawatts to the National Electric Grid of Pakistan [8] as shown in the figure below.

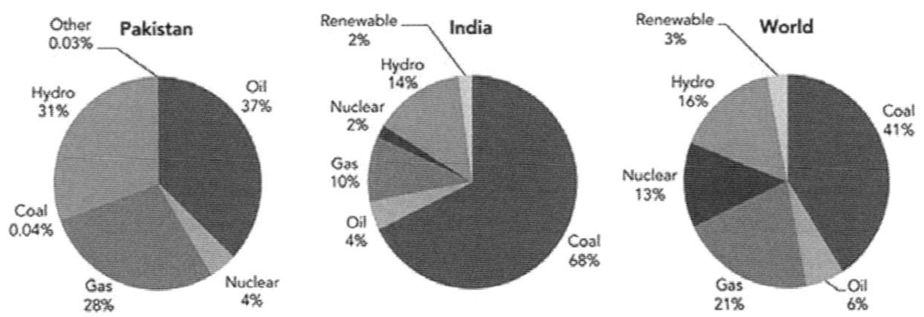

Figure 1. Type of sources used in power generation

Source: Kugelman Michael. [9].

To meet the rising energy demands [8] due to rapid population growth, the possible options are; (1) continue importing natural gas and fuel oil, (2) improve and identify alternative/renewable resources, (3) develop hydel projects or (4) exploit indigenous coal resources such as Thar coal which has a potential of generating 100,000MW per year for more than a century [10].

Now question is, will Thar coal mining be clean enough to support the sustainable development? Whether coal mining can help lighting the dark regions of Pakistan without further affecting environment? Some environmentalists argue that Thar coal should stay buried to have global temperature below 2°C, while others like Michael Kugleman a South Asian expert drawn a comparison [9] in figure 1 reflects Pakistan's reliance on coal for generating electricity. He echoed that Pakistan should be trying to access the coal from energy and economy point of view, he further added that "it is never too late for an energy-insecure country like Pakistan with huge amounts of reserves to start exploiting untouched coal, (therefore) it would be bad policy for both political and energy reasons to simply ignore nearly 200 billion tons of coal that potentially be exploited in Thar desert" [11] in an interview with Dawn news.

During recent decades, progress in the sphere of sustainable development has been made in many parts of the world. The resource-efficient *Green Economy* is one of the themes of international efforts to promote sustainable development in the 21st century. According to the World Bank, "better operations and the closure of clapped-out plants helped to push the average thermal efficiency of its coal-fired power stations from 31% in 2000 to 37% in 2010; America's remained flat, at 33%" [12].

However, there are several challenges that need to be addressed in the domestic coal industry, including better understanding of coal resources, improving coal extraction technology, and governance and regulation issues.

2. Methods and study area

The research methodologies include: a regional study; selecting the study region and information collection. For the purpose of this research, published reports and news articles, along with journal publications, were the major sources of information. The proposed research concepts used in this study is *circular economy*. The concept has been framed in response to growing concerns of resource scarcity and awareness that business is as usual unsustainable. Circular economy is a regenerative system in which an industrial shift towards resource input and waste, emission and leakage of energy can be achieved through long-lasting design, maintenance, repair, reuse, remanufacturing, and recycling, also it encourages the use of renewable energy to avoid the biosphere contamination through improved systems and business models.

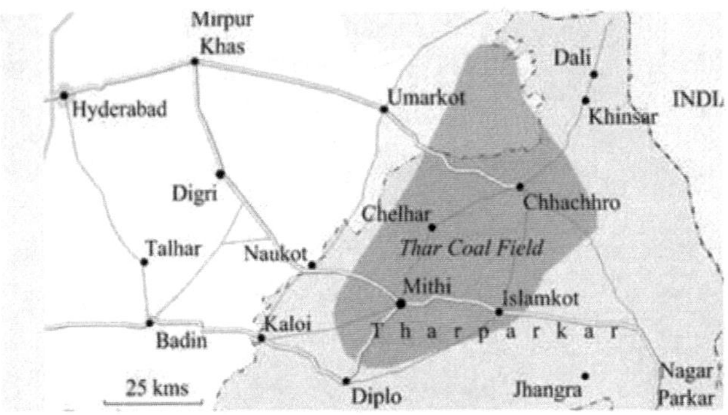

Figure 2. Location map of Thar coalfield.

Thar coalfield is located in the south-eastern part of Sindh as shown in figure 2. The first indication of coal beneath the sands of the Thar Desert was reported while drilling water wells by the British Overseas Development Agency (ODA) in coordination with the Sindh Arid Zone Development Authority (SAZDA), in 1988 [12]. Further exploration work was conducted by Geological Survey of Pakistan (GSP) and United States Geological Survey (USGS) under Coal Resources Exploration & Assessment Program (COALREAP) from 1989 to 1994 [13].

Pakistan possesses the 7th largest lignite resources in the world with nearly 200 billion tons of coal, mainly concentrated in Thar region (175 billion tons of lignite resources) [14]. Thar coalfield covers an area of approximately 9000 square kilometers. Cumulative seam thickness varies between 1.45 m and 42.6 m and the maximum thickness of an individual seam is 28.6 m [15].

3. Thar coal: quality and comparison

Sindh Engro Coal Mining Company (SECMC) engaged several renowned international organizations, including RWE Germany, SRK UK, Hagler Bailly Pakistan, Sino Coal China and NCGB China, all have conducted surveys to check the credibility of the Thar coal reserve and socio-economic impact assessments of the mining and power project [16]. The coal quality of Thar Block-II reported by Hagler Bailly Pakistan is lignite following Proximate Analysis. The raw coal accounts for 45.71% of moisture content, ash is 9.69%, volatile matters is 25.00% and 19.6% fixed carbon [16]. Environmental and Social Impact Assessment (ESIA) [17] shows that greenhouse gas emission from the Block-II power plant will be 4.9 million tons per year. Estimates using the IPCC Tier 1 methodology show a carbon emission of 101,000 kg per Tera-joule of heat input from lignite. A high content of aluminum oxide and silicon

IOP Conf. Series: Journal of Physics: Conf. Series **989** (2018) 012004 doi:10.1088/1742-6596/989/1/012004

dioxide, low ferric oxide and calcium oxide with minor amount of sulfur oxide and manganese oxide is been reported based on a chemical analysis of raw coal ash [18].

An average heating value of Thar coal (11.6 MJ/Kg) in contrast to heating value of similar kind of lignite coal reserves in other countries like, Rhineland lignite, Germany (8.9 MJ/kg) and Hungary lignite, Hungary (7.1 MJ/kg), makes Thar coalfield very feasible [18]. All above mentioned coal reserves account for thousands of megawatts of electricity against Thar coalfield's 0 MW (Thar Block-II will start generating 660MW of electricity by June 2018). The construction work on other blocks of Thar coalfield is also under process, according to A. Khan the managing director of SECMC, 4 blocks are under construction; Block-II is handled by SECMC, Block-IIIA handed to Cougar Energy (UK), Block-V is underground gasification project and Block-VI project is under Oracle Coalfields, PLC (UK) control, all four power projects will altogether add around 10,000 MW power into the national grid by 2020 [19].

Thar coalfield has an advantage of low-moderate sulfur over the world's average sulfur (2.42%) in lignite coal [20]. Oxides of sulfur are the most concerned emission pollutants from the regulatory stand point. Low sulfur content in Thar coal makes it acceptable for power generation without exceeding the emission standard for sulfur oxides [21, 22].

Apart from low Sulphur content, heating value and generation capacity of the reserve, the huge size of the reserve –175 billion tones makes a huge and adds a plus to the credibility of Thar coalfield, hence affirms it a more promising reserve that can serve the country for at least two centuries.

4. Economic impact on the region

Coal mining has a long legacy of providing jobs especially the places suffering from high poverty like Thar region. According to World Coal Association (WCA) report, even after decades, coal is still the backbone of modern electricity and currently supplies around 30% of primary energy and 41% of global electricity generation. It also predicts that the coal usage will rise over 50% by 2030, with developing countries responsible for 97% of this increase, primarily to meet improved electrification rates [23].

Pakistan's unreliable power supply is affecting the productive use of energy either directly or indirectly. The power shortage in the form of load-shedding annually cost Pakistan 14 billion Pakistani rupees (7% of GDP) [6]. Thar coal power project will mainly result in positive socioeconomic impact locally as well as regionally. With increased power generation, project (Block-II) only, will add 660 MW power to the national grid that will further be supplied to various sectors suffering from shortfall therefore, will have a positive impact on the country's economy through a better GDP. By providing additional job opportunities, project will directly impact the economy of the country with freshly created economic activities (employment and income) and indirectly by providing other opportunities like production, employment and other businesses that supply inputs to the project that will result in prosperity and stable incomes of locals [17].

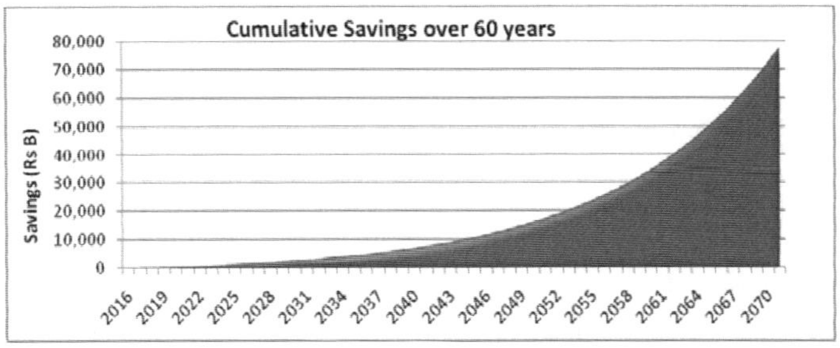

Figure 3. Net present value of savings from Thar coal project block-II [24].

According to Cambridge Energy Research Associates (CERA) estimates shown in figure 3, for oil prices Block-II will immensely benefit industrialists, domestic consumers and economy of the country by reducing the power generation cost and supplying uninterrupted electricity – net present value in Pakistani Rupees is 3.2 Trillion over a period of 60 years [24].

Low prices of electricity will help reviving the industry by providing more opportunities for investment – fallen down to 0.7% in recent years from 1.5% of GPD in 90s, and has resulted in forcible shut down of small and medium scale industries due to power crises. Also it will cut down the revenue spent on load shedding in the form of candles and kerosene [6].

Thar coal project will generate positive externalities not only in the form of hundreds of jobs but also by cumulative net savings and most importantly, regular and cost-effective energy supply needed to address the future energy mix of Pakistan.

5. Sustainable development in Thar coalfield

The coal industry around the world has made remarkable progress towards sustainable development to deal with social, environmental and economic aspects simultaneously (figure 4). The concept ensures the regional economic growth through sustainability goals, ecological and environmental protection in terms of efficient extraction of natural resources with minimum negative impact on environment. It also safeguards the social and community development of the mining area.

SECMC along with its sister concern Engro Powergen Thar Limited (EPTL) has been allocated Thar coal block-II for mining activities for two coal power plants each of 330MW capacity [16]. The adopted concept of circular economy in Thar coalfield helps improving the water quality, reducing greenhouse emission, hence successfully controlling water and air pollution.

Hagler Bailly Pakistan (ESIA) [17] has reported the sustainable development measures taken by SECMC in terms of water management out no risk to groundwater resource, the water will not be contaminated or pumped into the sea as wrongly believed, rather an effluent disposal storage pond is being constructed by the Sindh government with a treatment facility for recycling water. This treated water will be pumped to the power plant to increase the water supply.

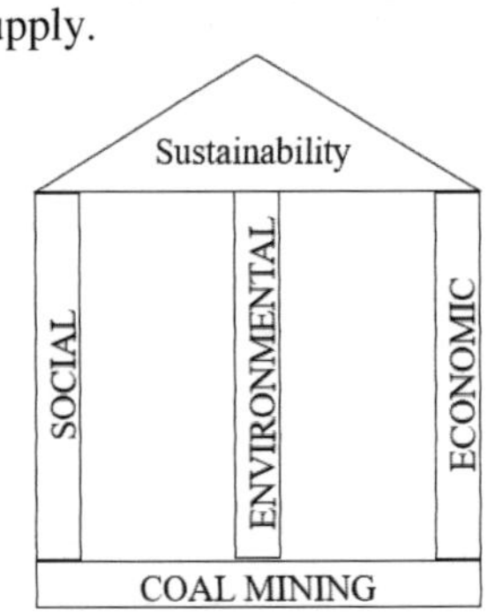

Figure 4. Coal mining sustainability model.

Regarding the coal combustion residual commonly called coal ash, only block-II project will produce 256,000 tons of ash, whereas the major part of it will be used to manufacture bricks and cement blocks [17]. Gypsum produced during the processing of coal, can be utilized in cement and marble industry. Another misconception about the large fugitive emissions that cause wind pollution has been rectified by the CEO of SECMC that the *dusk management* technique used reduces the adverse environmental impact and makes risk minimum and manageable [16].

The hazardous waste storage facility by SECMC at the site guarantees the storage and handling facility of the effluent wastes generated and emissions from the power plant are kept within National Environmental

Quality Standards (NEQS). An improved coal extraction and an advance processing technology has been adopted to minimize GHG emissions and environmental impacts such as the use of *Electrostatic Precipitator* in generators to minimize particulate emissions of fly ash and unburned carbon, whereas to control the emission of Sulfur dioxide (SO_2), limestone will be injected to circulating fluidized bed boiler (CFB) which converts Sulfur into gypsum and controlled emission of oxides of nitrogen will be provided by low NO_x burners [17].

6. Opportunity cost

Pakistan's reliance on mining for generating electricity till this day coal is almost nothing, therefore, there is a need to spread awareness among public about the threats coal mining poses regionally and globally.
Considering all sustainable measures, practically every mining activity especially coal mining somehow disturbs the natural environmental. Coal mining has mainly been criticized for its large amount of waste produced during coal mining, especially during the processing of coal. Seeing all under construction Thar coal power projects, only block-II will produce 256,000 tons of hazardous wastes annually. There are various risks involved regarding these wastes including the destruction of landscapes and habitats, emission of carbon and methane gas, acidification of the soil due to sulfur presents in sinkholes which leaves surface water and soil permanently hazardous for plants and animals.
The health impact of particulate matters in Thar coalfield due to open-pit mining is of greater importance than air pollution because, the level of PM_{10} and $PM_{2.5}$ received are above national standards which can cause diseases like black lung asthma, kidney disease and hypertension [25]. The emission of arsenic associated with the volatile organic matter of coal during combustion may cause environmental pollution in Thar region [21].

7. Suggestions

- Pakistan government needs to set a guiding ideology of national development that reflects the economic growth and supports the ecological sustainable development.
- There is a need to formulate number of laws, policies and strategies and their implementation is strictly needed, that can for example promote awareness about the Clean Production, circular economy production, climate change and other general laws of coal industry among the public.
- And lastly there is a serious need of public participation and corporation for their social responsibilities which is helpful in spreading environmental awareness about coal mines.

8. Conclusion

It is indubitable that Thar coalfield have potential to be financially profitable, technically reliable, environmentally sound and socially responsible, therefore, can meet the power needs of Pakistan for two centuries at least. And circular economy along with other sustainable techniques not only enhances the economic performance of Thar coalfield by regulating the needed supply of electricity which has paralyzed the economy of the Pakistan and by creating direct and indirect employment and business opportunities to alleviate poverty, but also reduces greenhouse gas emissions, controls water and air pollution and helps in addressing the environmental degradation problems. However the success of the coal industry and environmental development still largely depend upon the Pakistan governments' guiding ideology on development, strict enforcement of laws, policies, strategies, and public corporation and participation.

9. References

[1] Hancock S and Wolkersdorfer C, 2012, *Water Environment.* **31**, 147–158.
[2] Sivakumar R, Kannan D and Murugesan P 2014, *Resource Policy* **46**, 64–75.
[3] Ge J and Lei Y2013, *Resource Policy*, **38**, 278–287.
[4] Lu J and Lora- Wainwright A 2014, *World Development* **62**, 189–200.

[5] Energy use (kg of oil equivalent per capita) [Accessed May, 2017], *World Bank.* http://data.worldbank.org/indicator/EG.USE.ELEC.KH.PC
[6] Dawn News August 2016, 7 facts about Pakistan's energy crisis - and how you can help end it, *Dawn News.* (https://www.dawn.com/news/1275116)
[7] State of the economy: pulling back from the Abyss, 2010, Third Annual Report, *Institute of Public Policy* (IPP), Beacon-house National University Lahore.
[8] Power System Statistics 2014-2015 (Annual Report), Planning Power NTDC, *National Transmission and Dispatch Company* 40th Edition pp.a.
[9] Kugelman, Michael 2015, *Woodrow Wilson International Center for Scholars. Washington, DC.* Open report Pp. 157.
[10] Shah S and Khan Z 2013, *Middle-East Journal of Scientific Research* **15** (3): 327-331. https://www.idosi.org/mejsr/mejsr15(3)13/2.pdf
[11] Zofeen T. Ebrahim Feb 22, 2017, Economy vs environment: Thar coal and a test of Pakistan's priorities, *Dawn News*, https://www.dawn.com/news/1314947
[12] China and the environment: The east is not grey, August 10, 2013, *The Economist*, Beijing, Briefing.
[13] Fassett J and Durrani N 1994, Geology and Coal Resources of the Thar Coal Field, Sindh Province, Pakistan, *U.S. Geological Survey* Open-file Report 94-167.
[14] Singh, R.N., Atkins, A.S., Pathan, A.G 2010, *International journal of mining & environmental issues* **01**, 65–78.
[15] Pathan A, Singh R and Stace R 11-15 August, 2013, Geotechnical Assessment of Block VIII at Thar Coalfield, Pakistan, *23rd World Mining Congress,* Montreal, Canada.
[16] S.Shaikh, 21 March 2016, Thar coal – separating facts from fiction, Tribune the Express, https://tribune.com.pk/story/1069526/thar-coal-separating-facts-from-fiction/
[17] Thar block-II Power Project, Environmental and Social Impact Assessment Report 2014, *Hagler Bailly Pakistan.* Volume **1 of 2**, 6-1 – 6-24
[18] Choudry M, Nurgis Y, Sharif M, Mahmood A, Abbasi H 2010, *American Journal of Scientific Research* **11,** 92-102.
[19] Khan A, 2015b, Review of Pakistan Coal Industry Now & the Future, (http://pcc-usa.org/wp-content/uploads/2015/04/Thar-Coal-Opportunities.pdf).
[20] Bouska V and Pesek J 1999, *International Journal of Coal Geology* **40**, 211-235.
[21] Ali J, Kazi T G, Baig J A, Afridi H I, Arain M S, Naeem-Ullah, Arain S S and Siraj S 2015, *Journal of Geochemical Exploration* **159,** 227–233.
[22] Sarwar A, Khan M N and Azhar K F 2012, *Journal of Minerals and Materials Characterization and Engineering* **11**, 817-824.
[23] Coal: Energy for Sustainable Development, 2012, *World Coal Association*, p. 10.
[24] Khan A A, 2010, Thar coal: The Gateway to Energy Security of Pakistan, *Coal and Energy Development Department, Thar coal and Energy Board.* http://www.powershow.com/view/3b0277-NjYzY/Thar_Coal_Energy_Board_powerpoint_ppt_presentation
[25] Gautam S, Patra A K, Sahu S P and Hitch M. 2016, *International Journal of Mining, Reclamation and Environment*, http://dx.doi.org/10.1080/17480930.2016.1218110

Simultaneous Recovery of Carotenes and Tocols from Crude Palm Olein Using Ethyl Lactate and Ethanol

Yin Leng Kua[1], Suyin Gan[1], Andrew Morris[2] and Hoon Kiat Ng[3]

[1] Department of Chemical & Environmental Engineering, University of Nottingham Malaysia Campus, Jalan Broga, 43000 Semenyih, Malaysia
[2] Faculty of Science, University of Nottingham Malaysia Campus, Jalan Broga, 43000 Semenyih, Malaysia
[3] Department of Mechanical, Materials & Manufacturing Engineering, University of Nottingham Malaysia Campus, Jalan Broga, 43000 Semenyih, Malaysia
E-mail: kebx4kye@nottingham.edu.my

Abstract. This paper demonstrates the use of ethyl lactate and ethanol as green and safe solvents to extract phytonutrients such as carotenes and tocols from crude palm olein (CPO) before they are lost during oil refining process. The effects of mixing time (10-40 min), temperature (10-30°C) and proportion of CPO (20-60%) were studied in terms of the extraction of individual carotenes (α- and β-carotene) and tocols (α-tocopherol/T, α-, γ- and δ-tocotrienol/T3) in a temperature-controlled mixer-settler system. The optimal extraction conditions were found at 20°C, 10 min of mixing, 50% of CPO using 3:2 v/v ethyl lactate/ ethanol as the solvents. After four stages of extraction, 42.2% of carotenes, 86.7% of tocols and 44.4% of oil were recovered into an oil concentrate of 717.5 mg/L of carotenes and 1496.2 mg/L of tocols.

1. Introduction

Crude palm oil contains 500-700 ppm of carotenoids with approximately 35% α-carotene and 56% β-carotene [1]. They are important as a supplement of vitamin A as well as a natural antioxidant to fight against several degenerative diseases including cardiovascular diseases, cancers and macular degeneration [2]. In crude palm oil, there is also 600-1000 ppm of tocols (commonly known as vitamin E), with 21.3% of tocopherol (T) and 78.7% of tocotrienol (T3) [3]. As compared to T, T3 was reported to be more effective due to its unsaturated chain, which facilitates cell penetration and is highly antioxidative. T3 was reported to regulate cholesterol level, to prevent cancers, stroke and fats accumulation in liver [4]. T is commonly found in many other vegetable oils while palm and rice bran oil are among the richest natural sources of T3.

During physical oil refining process, carotenes are first partially removed by adsorption on activated bleaching earth, followed by high temperature steam deodorization which destroys the chromogenic properties of the remaining carotenes to produce a light yellow palm oil. Even though tocols are more thermally stable than carotenes, near to 50% of the tocols will be stripped off along with free fatty acids (FFA), sterols and squalene into palm fatty acid distillate (PFAD) during deodorization step. Thus, there is a need to recover these phytonutrients from crude palm oil before further refining process.

As both the extracted carotenes and tocols will be used as food fortifiers, the use of non-toxic, non-corrosive and non-carcinogenic solvents which are safe for human consumption is crucial. When most of the commonly used petrochemical solvents are known to pose certain degree of toxicity, ethyl lactate is a suitable candidate solvent because it is novel, green and safe. It is produced from the carbohydrate feedstocks from the corn and soybean industries and it presents naturally in foods such as wine, beer, chicken and fruits. It is also non-ozone depleting, non-hazardous air polluting and it is biodegradable into harmless compounds such as CO_2 and water. The US Environmental Protection Agency (USEPA) approved the solvent as a Significant New Alternatives Policy Program (SNAP) solvent while US Food and Drug Administration (USFDA) has approved its direct use in food and pharmaceutical products [5]. Ethyl lactate exerts polarity in the range of acetonitrile and n-hexane. It is capable to form intra- and inter-molecular hydrogen bonding. Also, it has the ability to form Van der Waals interactions in oils [6]. As a result, ethyl lactate can dissolve in both aqueous and hydrocarbon environments and it is capable to extract compounds of a wide range of polarity. Ethyl lactate has been reported to extract various nutraceutical compounds mostly from solid matrix [7]. Limited papers reported its potential to recover compounds directly from oil sample [8], [9].

The aim of this study is to investigate the potential of ethyl lactate and ethanol as safe and green solvents to extract carotenes and tocols from CPO via simple multistage mixing-settling process. Commercially, there is no technology available to date to recover these phytonutrients before palm oil refining process. Hence, this study provides the experimental results which can be used for future pilot scale recovery tests.

2. Material and methods

2.1. Samples and reagents

(S)-(-)-ethyl lactate (99% purity) and ethanol (99.5% purity) were obtained from Merck. Standards such as α-carotene (98% purity) was purchased from Fluka, β-carotene (99.4% purity) and α-T (100.9% purity) came from Calbiochem. α-, γ- and δ-T3 (97% purity) were bought from Davos Life Science (Singapore). Acetonitrile (HPLC grade) and dichloromethane (99.8% purity) were obtained from Labscan. 2-propanol (HPLC grade) came from Fisher while n-hexane (HPLC grade) from Merck. Crude palm oil (*Elaeis guineensis/tenera*) was centrifuged at 7500 rpm for 15 min in order to collect the upper olein phase for use in the extraction experiment.

2.2. Extraction process

In the temperature controlled mixer-settler system, a total volume of 250 mL solution composed of CPO, ethyl lactate and ethanol was prepared into the vessel. The proportion of CPO, ethyl lactate and ethanol vary for each run. The temperature of the solution was controlled at 10-30°C using an oil jacket controlled by the circulating bath (Lab Companion, Model RW-0525G). The solution was mixed at 360 rpm using a four-bladed stirrer for 10-40 min. After that, the mixture was left to settle for an hour with the temperature being controlled throughout. Lastly, the volume of the upper (oil) and lower (solvent) phases were measured and collected for analysis. Reversed- and normal-phase high performance liquid chromatography (HPLC) was employed to determine the concentration of carotenes and tocols, respectively, in both the oil and solvent phase. All the experiments were repeated and the average values were reported.

The process performance was evaluated mainly based on the enrichment factor, percentage recovery and percentage of oil recovery as defined by equation (1)-(4).

$$\text{Enrichment factor} = \frac{\text{mass of carotenes or tocols extracted (mg)/volume of oil extracted (mL)into solvent phase}}{\text{mass of carotenes or tocols (mg)/volume of oil (mL)present in feed}} \quad (1)$$

$$\text{Percentage recovery (\%)} = \frac{\text{mass of carotenes or tocols extracted (mg)}}{\text{mass of carotenes or tocols present in feed (mg)}} \times 100\% \quad (2)$$

$$\text{Total recovery (\%)} = \frac{\text{total mass of carotenes and tocols extracted (mg)}}{\text{total mass of carotenes and tocols present in feed (mg)}} \times 100\% \quad (3)$$

$$\text{Percentage oil recovery (\%)} = \frac{\text{volume of oil extracted (mL)}}{\text{volume of oil present in feed (mL)}} \times 100\% \quad (4)$$

2.3. High performance liquid chromatography (HPLC)

An Agilent 1260 Infinity Series HPLC system was used along with a quarternary pump, an autosampler, a variable wavelength detector (VWD) and ChemStation software for system control and data collection. Carotenes (α- and β-carotene) separation was carried out in a Purospher STAR RP-18 encapped column (5μm; 4.6 x 250 mm) maintained at 30°C. The mobile phase consists of 85% acetonitrile and 15% dichloromethane at 1.5 mL/min. The detector was set at 450 nm and the total run time was 25 min. Tocols (α-T, α-, γ- and δ-T3) separation was achieved in a Zorbax Rx-SIL column (5μm; 4.6 x 250 mm) maintained at 30°C. The mobile phase consists of 99% of n-hexane and 1% of 2-propanol at 0.8 mL/min. The detector was set at 292 nm and the total run time was 15 min.

3. Results and discussion

3.1. The use of ethanol as co-solvent

Table 1, 2 and 3 present the results of extraction when operated at 20°C, 25 min of mixing and 50% of CPO at a total capacity of 250 mL. When ethyl lactate was used as the sole solvent, the total carotenes were concentrated by 0.92 while the total tocols were concentrated by 2.25 (table 1). Tocols exert greater affinity towards ethyl lactate because they are more polar than carotenes due to the presence of hydroxyl group and unsaturated side chain. Even though ethyl lactate exerts polarity in the range of acetonitrile and hexane, it tends towards polar solvent. Therefore, compounds of higher polarity are more readily extracted. Similarly, the enrichment factor improved from 1.69 to 3.02 in the order of α-T, α-T3, γ-T3 and δ-T3 with increasing polarity. Since β-carotene was more concentrated than α-carotene, β-carotene is more polar than α-carotene. A total recovery of 13.5% was achieved including both carotenes (7.6%) and tocols (18.6%) in a single stage system. The oil recovery was 8.2%.

Table 1. The concentration of each carotenes and tocols in feed, solvent and oil phase (after solvent removal) when 100% of ethyl lactate was used as the solvent.

Concentration (mg/L)	Feed	Solvent	Oil	Enrichment factor
α-carotene	311.4	286.2	313.4	0.91
β-carotene	461.2	426.9	463.9	0.92
Total carotenes	**772.6**	**713.1**	**777.3**	**0.92**
α-T	242.9	409.5	227.8	1.69
α-T3	222.4	465.5	200.3	2.06
γ-T3	326.2	859.7	277.9	2.61
δ-T3	85.4	261.3	69.5	3.02
Total tocols	**877.0**	**1996.0**	**775.4**	**2.25**

After mixing and settling of a solution made up of 125 mL of CPO and 125 mL of ethyl lactate, 78 mL of the solvent phase was collected. It was only 62.5% of the initial volume of ethyl lactate introduced into the vessel. Ethyl lactate diffused and retained in the oil phase due to the formation of Van der Waals interactions. In order to create a better two-phase system with stronger immiscibility, ethanol, which is strongly polar, green and safe, was introduced as a co-solvent. The addition of ethanol aims to refrain ethyl lactate from diffusing into the oil phase. [10] reported that more β-carotene was extracted from carrots powder with increasing amount of ethanol as a co-solvent in ethyl lactate. Therefore, ethanol was chosen to improve the overall performance. 40% of ethanol was used with 60% of ethyl lactate as the mixed solvents because ethanol was found to form another individual top phase as the volume goes beyond 50%. This was due to the density differences of ethanol (789 kg/m^3), CPO (888 kg/m^3) and ethyl lactate (1034 kg/m^3). In the presence of 40% ethanol, the volume of solvent phase improved to 89 mL, which was 71.2% from the initial volume (125 mL) of solvents added. After ethanol addition, the enrichment factor of carotenes and tocols improved to 0.96 and 2.68, respectively (table 2). Again, the extraction of tocols was much better due to their polarity difference. Similar results were reported by [11] as more T dissolved in ethanol than carotenoids. Even though the oil recovery increased to 11.0%, the total recovery improved to 21.4% with carotenes and tocols recovery increased to 10.5% and 29.3%, respectively

Table 2. The concentration of each carotenes and tocols in feed, solvent and oil phase (after solvent removal) when 40% ethanol + 60% ethyl lactate (3:2 v/v ethyl lactate/ethanol) was used as the premixed solvents.

Concentration (mg/L)	Feed	Solvent	Oil	Enrichment factor
α-carotene	243.8	199.9	249.3	0.81
β-carotene	402.1	419.3	400.0	1.04
Total carotenes	**645.9**	**619.2**	**649.2**	**0.96**
α-T	239.5	531.2	203.5	2.21
α-T3	271.8	657.0	224.3	2.42
γ-T3	363.6	1087.0	274.4	2.99
δ-T3	89.8	306.2	63.1	3.41
Total tocols	**964.6**	**2581.5**	**765.5**	**2.68**

In table 3, 100% ethanol was used as the solvent for extraction. The enrichment factor of tocols was further improved to 2.76 but it dropped drastically to 0.56 for carotenes. Since ethanol is a strongly polar solvent, it is not suitable to be used alone to recover carotenes, which are non-polar in nature. 6.1% of carotenes and 30.2% of tocols were recovered with a total recovery of 20.1%. The oil recovery was 10.7%. Apart from carotenes, the extraction performance of 100% ethanol was comparable to the results as obtained by using 40% of ethanol and 60% of ethyl lactate in table 2. Therefore, 40% ethanol and 60% ethyl lactate (equivalent to 3:2 v/v ethyl lactate/ethanol) were used in the subsequent extraction experiments to improve the yield.

Table 3. The concentration of each carotenes and tocols in feed, solvent and oil phase (after solvent removal) when 100% of ethanol was used as the solvent.

Concentration (mg/L)	Feed	Solvent	Oil	Enrichment factor
α-carotene	275.4	146.5	291.0	0.53
β-carotene	342.4	199.4	359.0	0.59
Total carotenes	**617.7**	**345.9**	**650.0**	**0.56**
α-T	212.0	474.5	178.5	2.21
α-T3	200.3	497.2	162.1	2.53
γ-T3	330.6	1017.1	245.3	3.05
δ-T3	87.7	320.6	59.7	3.66
Total tocols	**830.5**	**2309.3**	**645.6**	**2.76**

3.2. The effect of mixing time

Table 4 shows the enrichment factors and recoveries as obtained after 10, 25 and 40 min of mixing at 20°C and 50% of CPO using 3:2 v/v ethyl lactate/ethanol as the solvents. As the mixing time increased from 10 to 25 min, the enrichment factor and recovery for both carotenes and tocols were improved. A longer period of agitation induced higher chances of bringing the solvent and CPO into direct contact and hence, there was improved mass transfer into the solvent phase. Further increase in the mixing time to 40 min reduced both the enrichment factors and recoveries. The formation of stable emulsion after prolonged mixing might affect the extractability negatively [12]. Also, competitive extraction might occur when other compounds such as FFA, glycolipids, sterols and squalene [13] in CPO were co-extracted. As a result, carotenes and tocols back-diffused into the oil phase after prolonged period of mixing.

The improvement of enrichment factors and recoveries as mixing time increased by 15 min from 10 min were not pronounced. Mixing time had little effect on the process performance. Therefore, further increase of the mixing time by 1.5 times was not necessary. In the industry, multistage extraction is preferred to improve the overall extraction yield. A short mixing or operating time is desirable to reduce the overall operational takt time. Thus, the subsequent experiments were carried out at 10 min of mixing.

Table 4. The enrichment factor and recovery of carotenes, tocols and oil when operated at mixing time of 10, 25 and 40 min.

Mixing time (min)		10	25	40
Enrichment factor	Carotenes	0.88	0.96	0.77
	Tocols	2.56	2.68	2.26
Recovery (%)	Carotenes	10.1	10.5	9.0
	Tocols	28.0	29.3	26.1
	Total	20.3	21.8	18.6
Oil recovery (%)		11.2	11.2	11.7

3.3. The effect of temperature

Figure 1 and 2 illustrate the results of the enrichment factors and recoveries when operated at 10, 20 and 30°C after 10 min mixing of 50% of CPO with 3:2 v/v ethyl lactate/ethanol. Temperature as low as 15.2°C was reported to be superior to extract α-T from olive oil [9]. Lower operating temperature is preferred to avoid thermal decomposition of heat-sensitive compounds such as carotenes and tocols at high temperature. However, higher temperature system provides more energy, the kinetics will be higher to induce mass transport. In addition, solvents become less viscous for diffusion at higher temperature and hence, the diffusivity or rate of diffusion is expected to be enhanced.

As the operating temperature of current system reduced from 20 to 10°C, there were minimal effects on the enrichment factors and recoveries. There was slight reduction for carotenes as the kinetic of mass transport was slower at 10°C. Longer mixing time might be required to improve the yield. In contrast, the performance of tocols extraction remained almost unchanged. Thus, additional cooling below normal atmospheric temperature was not necessary.

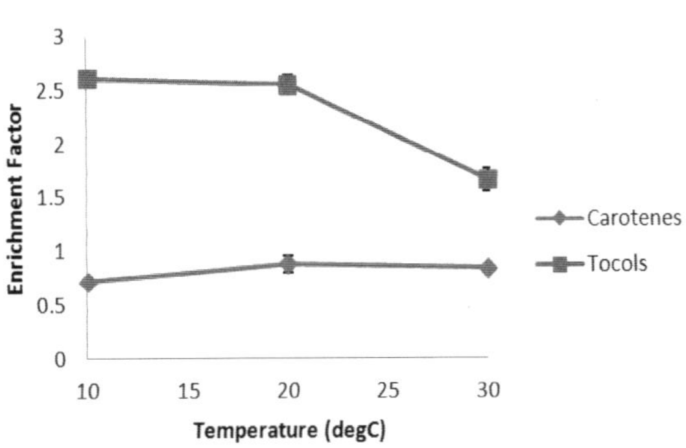

Figure 1. The enrichment factor of carotenes and tocols at 10, 20 and 30°C.

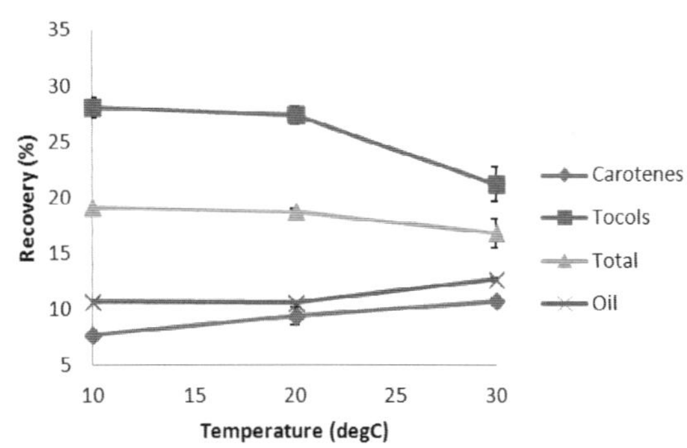

Figure 2. The percentage recovery of carotenes, tocols and oil at 10, 20 and 30°C.

As the temperature increased further to 30°C, the enrichment factor and recovery of carotenes remained almost unchanged while they reduced drastically for tocols. Under such conditions from 10 to 30°C, temperature was not a determining factor to affect the extraction of carotenes. The solvents content in the oil phase ranged from 31.2, 32.2 and 42.0% as temperature increased from 10 to 30°C. The oil recovery was also getting higher at 30°C. With increasing temperature, more solvent diffused into the oil phase and vice versa. As a result, the selectivity reduced as more oil but less tocols were extracted at higher temperature. Similar results were reported by [14] whereby selectivity reduced at higher temperature. As there was more oil extracted at higher temperature with no additional carotenes and tocols being recovered, 20°C was chosen as the optimal temperature with best results.

3.4. The effect of the volume percentage of CPO

Figure 3 and 4 show the enrichment factor and recovery of carotenes, tocols and oil at 20, 30, 40, 50 and 60% of CPO when operate at 20°C and 10 min of mixing time using 3:2 v/v ethyl lactate/ethanol as the solvents. Referring to figure 3, 50% of CPO gave the highest enrichment factors for carotenes and tocols at 0.88 and 2.56, respectively. As the volume of CPO was less than 50%, there was lesser amount of carotenes and tocols available for extraction and hence, the driving force (concentration difference) for mass transfer would be lower. As opposed to tocols, the enrichment factor of carotenes remained almost constant near at 0.9. This implied that it was the maximum enrichment factor which could be achieved using 3:2 v/v of ethyl lactate/ethanol as the solvents. Any further increase in the amount of CPO will not improve the enrichment factor of carotenes. As the volume of CPO increased to 60%, the amount of carotenes and tocols increased along with other minor compounds such as FFA, glycolipids, squalene aand sterols. Thus, the enrichment factor became lower largely due to competitive co-extraction.

The maximum recovery occurred at the lowest volume percentage of CPO at 20% as shown in figure 4. From 60% of CPO, the recovery of carotenes and tocols increased by 33.2 and 62.4%, respectively. The total recovery at 20% of CPO reached 55.1% while the recoveries of 36.5% carotenes and 71.6% tocols were achieved. The oil recovery increased along with phytonutrients recovery from 4.1 to 38.8% as the volume of CPO reduced. Since the volume of solvents increased when CPO reduced, more compounds including carotenes, tocols and oil were extracted leading to high recovery. Carotenes and tocols were minor components while oil (triglycerides) present in bulk. Therefore, the concentration difference was maintained for oil even at the lowest volume of CPO. As a result, more oil was extracted causing low enrichment factor with reducing volume of CPO as can be seen in figure 3.

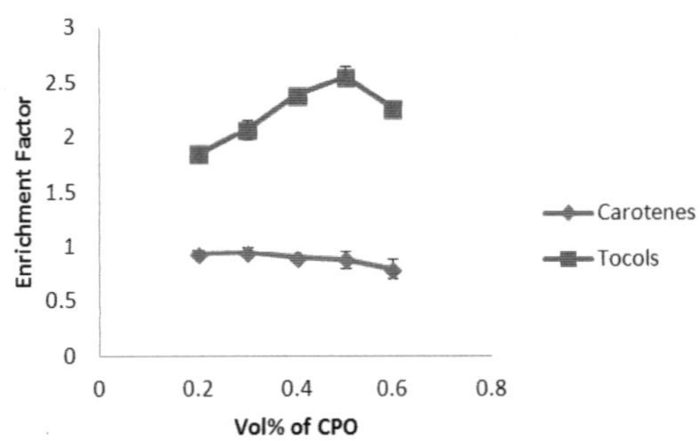

Figure 3. The enrichment factor of carotenes and tocols at 20, 30, 40, 50 and 60% of CPO.

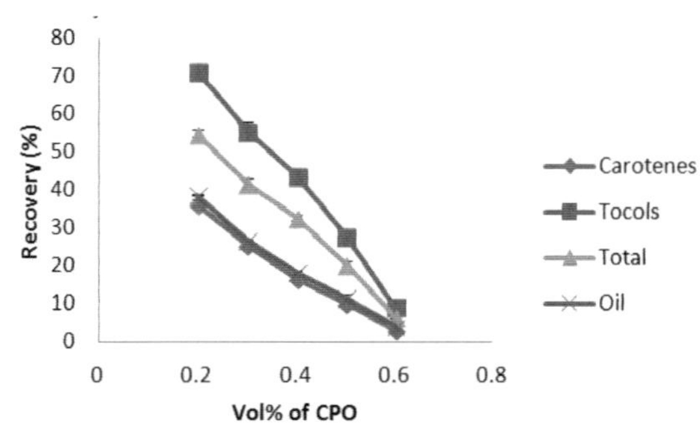

Figure 4. The percentage recovery of carotenes, tocols and oil at 20, 30, 40, 50 and 60 % of CPO.

As the remaining treated oil phase is returned to the oil refineries, the increase of the subsequent oil throughput due to solvent addition and the recovery of oil into solvent phase are not wanted. As the volume of CPO increased, both oil throughput and solvent content in the oil phase increased as illustrated in figure 5. As the volume of CPO increased, less oil diffused into the solvent phase, but more solvents diffused into the oil phase. The oil throughput should be near to unity while the oil recovery should be minimized. Hence, the optimal point was controlled at 50% of CPO when the oil throughput is 1.25 with 32.5% of solvents in the oil phase while 11.6% of oil was recovered into the solvent phase. Subsequent solvent removal from the oil phase could be carried out before conventional oil refining process. However, the presence of these solvents in the oil does not affect the safety of the palm oil because they are both food grade and safe for human consumption. At 50% of CPO, the highest enrichment factors were achieved at a total recovery of 20.3%. Test at 10% of CPO was also investigated, but was not reported here because the oil phase became too little (~11.5 mL) to be measured accurately. In addition, the overall operation takt time would increase as only 10% of CPO could be treated in a single run with lower enrichment factor for tocols as more oil was recovered. The optimum with the best extraction performance was found at 50% of CPO, 10 min of mixing at 20°C using 3:2 v/v ethyl lactate/ethanol as the solvents.

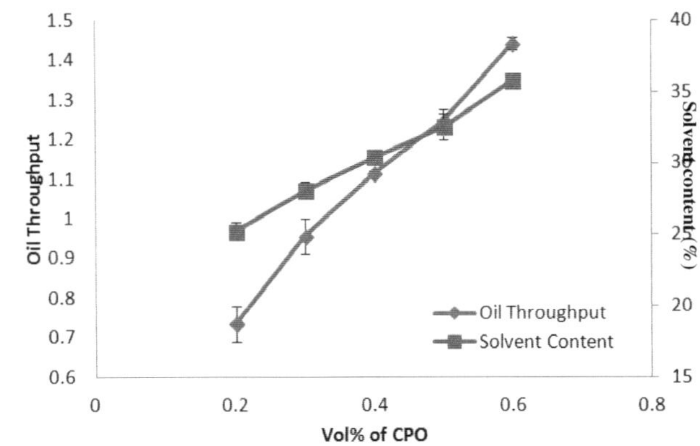

Figure 5. The oil throughput and solvent content in the oil phase from 20 to 60% of CPO.

3.5. Multistage extraction

At 20°C, 10 min of mixing and 50% of CPO using 3:2 v/v ethyl lactate/ethanol as the solvents, four stages of mixing-settling were carried out. Upon completion of a stage, the solvent phase at the bottom was withdrawn, measured and analyzed. Fresh premixed solvents were then added for another cycle of mixing-settling. Table 5 is a compilation of the enrichment factors and recoveries in each stage while table 6 compiles the cumulative results, including the previous fraction(s) collected. The cumulative enrichment factor for carotenes increased while it decreased for tocols along the stages. The improvement of the cumulative enrichment factor for carotenes was not obvious because the enrichment factor remained close to unity in each stage. Conversely, the cumulative enrichment factor for tocols reduced in subsequent stages. This was because the concentration of tocols reduced after each extraction cycle and hence, the concentration of the extracted tocols reduced from 2074.1 to 887.8 mg/L even though higher separation (enrichment factor) was achieved.

Table 5. The enrichment factor and percentage recovery of carotenes, tocols and oil in each stage.

Stage		1	2	3	4
Enrichment factor	Carotenes	0.82	0.90	0.98	1.09
	Tocols	2.73	2.68	2.94	3.37
Recovery (%)	Carotenes	10.1	15.3	13.2	12.7
	Tocols	33.5	45.6	39.4	39.3
	Total	21.8	28.3	22.1	19.4
Oil recovery (%)		12.3	17.1	13.5	11.7

Table 6. The cumulative enrichment factor and percentage recovery of carotenes, tocols and oil over stages.

Stage		1	2	3	4
Enrichment factor	Carotenes	0.82	0.88	0.92	0.96
	Tocols	2.73	2.34	2.11	1.96
Recovery (%)	Carotenes	10.1	23.8	33.9	42.2
	Tocols	33.5	63.8	78.1	86.7
	Total	21.8	44.0	56.2	64.6
Oil recovery (%)		12.3	27.3	37.0	44.4

After these four cycles, four parts of solvents were utilized for extraction per part of CPO. This was similar to the solvent consumption at 20% of CPO as presented earlier in section 3.4. It was found that with the same amount of solvent being consumed per part of CPO, multistage extraction achieved higher enrichment factors and recoveries. When single stage was carried out, the percentage oil recovery at 38.6% was only slightly lower as compared to multistage at 44.4%. Therefore, multistage extraction performed better than single stage system when the same amount of solvent was used.

4. Conclusions

The process of carotenes and tocols recovery was optimized at 20°C, 50% of CPO, 10 min of mixing using 3:2 v/v/ ethyl lactate/ethanol as the solvents. The operation was fast, simple, green and safe which can be retrofitted the existing palm oil refining line as a cheaper alternative of red palm oil production. It was carried out in a temperature-controlled mixer-settler unit to recover 42.2% of carotenes and 86.7% of tocols in CPO enriched by 0.96 and 1.96 for carotenes and tocols, respectively, after four extraction cycles.

5. References

[1] Choo Y M, Ong A S H, Goh S H and Khor H T 1990 Transesterification of fats and oils *UK Patent No 2188057.*

[2] Olson J A 1999 *Modern Nutrition in Health and Disease: Carotenoids* 9th Edition, ed Shils M E, Olson J A et al. (Baltimore: Willians & Wilkins) pp 525-41.

[3] Ong A S H 1993 *Vitamin E in health and disease: Natural sources of tocotrienols*, ed Packer L and Fuchs J (New York: Dekker) pp 3-8.

[4] Gopalan Y, Shuaib I L, Mogosso E, Ansari M A A, Bakar M R, Wong J W, Khan N A K, Liong W C, Sundram K, Ng B H et al. 2014 Clinical investigation of the protective effects of palm vitamin E tocotrienols on brain white matter *Stroke* **45(5)** pp 1422-8.

[5] Food and Drug Administration Department of Health and Human Services 2005 *The Code of Federal Regulation* Title 21 Food and Drugs (Washington: Government Printing Office).

[6] Aparicio S, Hallajian S, Alcalde R, Garcia B and Leal J M 2008 Liquid structure of ethyl lactate, pure and water mixed, as seen by dielectric spectroscopy, solvatochromic and thermophysical studies *Chemical Physics Letters* **454** pp 49-55.

[7] Kua Y L, Gan S, Morris A and Ng H K (2016) Ethyl lactate as a potential green solvent to extract hydrophilic (polar) and lipophilic (non-polar) phytonutrients simultaneously from fruit and vegetable by-products *Sustainable Chemistry and Pharmacy* **4** pp 21-31.

[8] Hernandez E J, Luna P, Stateva R P, Visak V N, Reglero G and Fornari T 2011 Liquid-liquid phase transition of mixtures comprising squalene, olive oil and ethyl lactate: Application to recover squalene from oil deodorizer distillates *Journal of Chemical and Engineering Data* **56** pp 2148-52.

[9] Vicente G, Paiva A, Fornari T and Visak V N 2011 Liquid-liquid equilibria for separation of tocopherol from olive oil using ethyl lactate *Chemical Engineering Journal* **172** pp 879-84.

[10] Ishida B K and Chapman M H 2009 Carotenoid extraction from plants using a novel, environmentally friendly solvent *Journal of Agricultural and Food Chemistry* **57** pp 1051-9.

[11] Gonçalves C B, Filho P A P and Meirelles A J A 2007 Partition of nutraceutical compounds in deacidification of palm oil by solvent extraction *Journal of Food Engineering* **81** pp 21-6.

[12] Ahmad A L, Chan C Y, Abd Shukor S R and Mashitah M D 2008 Recovery of oil and carotenes from palm oil mill effluent (POME) *Chemical Engineering Journal* **141** pp 383-6.

[13] Choo Y M and Ab Gapor M T 1990 Minor components from palm methyl esters *Palm Oil Developments* **14** pp 39-44.

[14] Baharin B S, Abd Rahman K, Abd Karim M I, Oyaizu T, Tanaka K, Tanaka Y and Takagi S 1998 Preparation of palm carotene from crude palm oil by adsorption chromatography with a synthetic polymer adsorbent *Journal of American Oil Chemists' Society* **75(3)** pp 399-404.

Acknowledgements

The crude palm oil used in this work was kindly provided by Havys Oil Palm Sdn. Bhd. (Semenyih). This work was supported by the Ministry of Science, Technology and Innovation (MOSTI), Malaysia under eScience Fund [grant number 02-02-12-SF0225]. The Faculty of Engineering at the University of Nottingham Malaysia Campus is also acknowledged for its support towards the project.

A Review of Water Reclamation Research in China Urban Landscape Design and Planning Practice

Wei Gan[1] and Tianran Zeng[2]
[1] Associate Professor, Huazhong University of Science and Technology, Wuhan, China
[2] Master, Georgia Institute of Technology, Georgia, United States

Abstract. With the continuously growing demand for better living environment, more and more attention and efforts have been paid to the improvement of urban landscape. However, the expansion of green area and water features are at the cost of high consumption of water resources, which has become prominent problems in cities that suffer from water shortage. At the same time, with the water shortage and water environment deterioration problems that shared globally, water conservation has become an inevitable choice to achieve sustainable social development. Urban landscape is not simply a consuming body of water resources, but also are of water-saving potential and able to perform the function of water storage. Thus, recycling the limited water resources becomes a challenge for every landscape designer. This paper is intended to overview the existing effort of reclaimed water recycle research in China landscape designing fields, and raise recommendations for future research and development.

1. Background

The concept of reclaimed water is usually relative to tap water and the waste water (discharged into the pipeline of sewage), refers to the wastewater appropriately treated to meet certain water quality standards, recycled and qualified for certain domestic (not drinkable) and industrial usage. Reclaimed water reuse can not only greatly slow down pace of the natural surface water and groundwater consumption, alleviate the shortage of water resources, ecological and geological problems, but also reduce the discharge of wastewater, effectively inhibit the environmental pollution of wastewater discharge. Therefore, the study of water reuse is of practical significance. This section is to introduce the existing international effort of water reclamation and the water recycle research in China landscape design and planning fields.

2. Overview of water recycling research for landscape design

2.1. Water recycling research for landscape design at international scale

The idea of recycling reclaimed water had already been widely in 1930s, and those reclaimed water were mostly used for urban landscape scenic water [1]. Internationally, the study of water reuse has been developed relatively well in some developed countries, where supportive policies have been introduced as well.

The United States has been one of the first counties to implement water reuse. According to the US Environmental Protection Agency, the reclaimed water shall not be used as drinking water and is mainly limited to landscape, ecological and industrial use. In 1923, the first large-scale sewage water recycling project for landscape usage was established in San Francisco, which filled the lake in parks with reclaimed water and had aroused significant impact over the world. Since then, more than 300 cities in the United States have establish sewage treatment projects, the recycling of reclaimed water has been further improved, greatly alleviated the contradiction between water and achieved the efficient use of water resources to some extent [2].

In Australia, a project located in the residential area around the Moxon Lake in Adelaide has been established to collect the sewage water and the rainwater from rooftop, and to meet the supply of Monson Lake water, residential water and landscape land irrigation after reclamation treatment.

In Japan, with its complex terrain and short rivers, the shortage of fresh water resources has urged the promotion of water reuse research in Japan. Currently, about 65% of the total population in Japan are using reclaimed water, and the national average annual use of circulating water has amounted to 1.5 × 108m3. A guideline for the reuse of water resources was introduced by Tokyo government in 1984, requiring the use of recycled water in buildings where construction area exceeded 3 x 104 m2 or daily water consumption reached 100 m3. Reclaimed water has been widely used in four aspects in Japan: household water reuse system, regional water reuse system, regional water reuse system and industrial water use system [3].

In 2000, the Jurong Island Industrial Park of Singapore, a set of urban sewage depth treatment device was constructed, with the scale of reclaimed water production of 30,000 m3 · d-1, and the treated water was mainly used for landscape water and fire water system (Chenxiang Xia,2007).

South Africa, one of the countries with the most serious water shortage problem, has been relatively more mature in its research on water treatment and reuse. About 22% of the country's water resources are recycled water. South Africa has established a dedicated sewage treatment agency to reclaim waste water and deliver to various areas for reuse, to meet the country's water needs [1].

European countries, Saudi Arabia, Mexico and other countries are also actively studying the methods and forms of reclaimed water reuse, and has achieved a series of results for the recycling of water resources to contribute [1].

2.2. *Water recycling research for landscape design of china*

Compared with foreign countries, China's water shortage has been paid attention relatively later, and the water reuse technology related research and development were first carried out in late 1980s. From 1986 onwards, the China government included the urban sewage resources exploitation in the "Seventh-Five" "Eighth-Five" and "Ninth Five-Year Plan", and gradually accept and explore the water reuse in the urban landscape fields.

The water pollution control and urban sewage technology were included in the national "Seventh-Five" research program(1986 - 1990). Tianjin, Beijing, Dalian, Taiyuan, Xi'an and other water-saving cities have carried out on the sewage regeneration technology and technology research, among which, Tianjin developed a "Tianjin landscape river water quality standards", beginning the attempt of applying Class two treated water that treated by Jizhuangzi sewage treatment plant for the landscape supplement [4].

Later, the national "eight five" research program has included the sewage resources and sewage reuse into the major research projects. During the "Eighth Five-Year Plan" period, Beijing Gaobeidian sewage treatment plant, the North River sewage treatment plant, Jiuxianqiao sewage treatment plant, Shijiazhuang West sewage treatment plant, Shandong Tai'an City sewage treatment plants and some other recycle water plants that were intended for urban landscape water usage had been completed and put into production [4]. After being recycled and treated to certain standard of purification, part of the water would be supplied to urban landscape and river water body. Among them, the Beijing Beixiaohe sewage treatment supplied 20000m3 / d of recycled water to greening, river water supplement; Shenzhen Glacier City Sewage Treatment Plant secondary effluent 20000m3 / d reclaimed water for three-level treatment for park greening and landscape water [4]. Supplying reclaimed water as a city landscape ecological water source revealed that China began to gradually accept and explore the use of reclaimed water in urban landscape [5].

In the "Ninth Five-Year Plan" (1996-2000), China began to study the key technologies of sewage treatment and water industry. During the period of the 10th Five-Year Plan (2001-2005), the water reuse was formally written into the document. The research on the utilization technology and demonstration projects of urban sewage resources was also carried out in a comprehensive manner. Beijing, Tianjin, Qingdao and other cities have set up a batch of landscape water reuse demonstration projects to demonstrate the reuse of recycled water in China's landscape water engineering practice.
Subsequently, in the "Eleventh Five-Year" plan (2006-2010), the water reuse and urban sewage treatment works were listed as one of the nine key projects supported by the government. During the 12th Five-Year Plan period, China has further promoted the recycling of sewage in the landscape. Beijing, Dalian and other cities have increased their use of reclaimed water every year. The reclaimed water has been used to remedy the drying rivers and rebuild the river landscape.
Currently, many cities in China has started to rectify rivers and water bodies heavily polluted which due to various improper maintenance. However, plenty of available water resources were wasted and thus led to harsher environmental problems.
Blindly using clean water resources to replace the polluted rivers and water bodies without in-depth study of how to better maintain the purification of water would eventually lead to gradually deteriorating water quality and require more water for replacement, which would undoubtedly be a great waste of existing resources. In recent years, Beijing, Shenzhen and other cities has been dealing with the reclaimed sewage water after further purification and put it into the use as urban landscape water, which has played a positive role in promoting China's urban landscape water recycling research.

3. Reclaimed water quality standards and remediation methods

3.1. Reclaimed Water Quality Standards

The main control indicators of recycle water quality are pH value, color, turbidity, BOD_5, CODcr, dissolved oxygen and phosphorus, nitrogen, etc. These indicators reflect the evaluation of equipment corrosion, sensory comfort, water safety and water quality maintaining and so on. Among several standards for water quality published by China government, The Reuse of Urban Recycling Water. Water Quality Standard for Scenic Environment Use (GB / T18921-2002) has been the standard for the assessment of landscape water quality, which suggests water reuse for urban landscape scenic water quality indicators in table 1 as followed:

Table 1. Recommended Water Quality Standards for Urban Landscape Reuse

Index	Indicator	Recycle Water Quality Standards
1	Basic requirements	No floating matter, No unpleasant smell and flavor
2	PH value	6.0~9.0
3	Chroma (degrees)	≤30
4	Turbidity (NTU)	≤10
5	Suspended solids (mg / L)	≤15
6	Soluble solids (mg / L)	≤1000
7	BOD5（mg/L）	≤10
8	CODcr（mg/L）	≤50
9	Total phosphorus (calculated as P) (mg / L)	≤1.0
10	Total nitrogen (mg / L)	≤15
11	Ammonia (in N) (mg / L)	≤5
12	Residual chlorine (mg / L)	≥0.05
13	Dissolved Oxygen (mg / L)	≥1.5
14	Anionic Surfactant (mg / L)	≤0.5
15	Fecal coliform (a / L)	≤10000

IOP Conf. Series: Journal of Physics: Conf. Series **989** (2018) 012006 doi:10.1088/1742-6596/989/1/012006

4. Remediation methods

4.1. Physical and chemistry remediation

Among several factors that impact the sustainability of water resources and runoff pollution, water pollution is particularly prominent and cause the waste of water resources in urban landscape construction process. For scenic water that of strong liquidity, purification facilities, such as the filter dam, the filter sand tank, etc. are applicable in the early stage of design, to maintain water quality and reduce the turbidity naturally with the natural flow of water.

However, for artificial water with poor fluidity, slow flow, small size, and were mostly closed, the water body itself is often of poor self-purification capacity, vulnerable to secondary pollution, and would indirectly affect the visual effects of the landscape. Thus, to control the water resource exogenously, restore water body physically while adding chemical agents to remove water pollutants are commonly used as solution to achieve water purification and maintenance.

4.2. Bioremediation methods

Aquatic plants have been playing a certain role in the purification of sewage, they can absorb and filter contaminants in the water through roots, thus cultivation of aquatic plants in landscape water is an effective way to repair the reclaimed water quality. The ecological repair technology has its unique advantage. The cost of this repair technology and adverse effects on natural environment are relatively low, and a reasonable design with aquatic plants can also create a rich aquatic landscape and enhance the environment. Nevertheless, the water purification function of aquatic plants produces the desired effect relatively slower compared with other artificial technology, and would also be impacted by seasons and temperatures.

5. Problems and suggestions for improvement of reclaimed water reuse in china

5.1. Existing problems of water reuse in china

5.1.1. Public awareness. Public awareness is the prerequisite and basis for supporting water recycle for landscape use. In southern China, especially in some coastal cities, it is generally believed that water resources are rich, thus lead to weak consciousness of water recycle and water resources, and a considerable part of the public has not yet formed the concept of water reuse. Without public awareness and acceptance of water reuse, even in those northern cities that short of water resources, the reclamation and recycle of water difficult would be hard to achieve. Moreover, the lack of confidence in the safety of water quality from the public also restricts the domestic water reuse for landscape development.

5.1.2. Lack of policy support. Using reclaimed water for urban landscape construction is conducive to the sustainable development of society, and is of no doubt should be given policy support and concession by the Government. To attract stakeholders who reuse water for landscape construction, the use of water should be more cost-effective than using tap water. For the water production enterprises, currently, the production of water that meet the quality requirements request additional costs, thus, the preferential policies from government would be preferable for those enterprises. At present, the policies to enhance water reuse in China urban landscape design field are still almost blank and in urgent need of development.

6. Suggestions of improvements

6.1. Public awareness.

Water reuse plays an active role in social sustainable development and environmental protection. However, it is not sufficiently concerned in southern China, where water resource is currently abundant. Meanwhile, urban landscape water reuse is facing the same problem. Therefore, it is necessary to make the public aware of the social, economic and ecological significance of the reuse of reclaimed water in the landscape and to enhance its advocacy. The government needs to popularize the water reuse technology, landscape water quality standards, and introduce domestic and foreign successful landscape reclaimed water reuse examples to educate the public, make them willing to accept and trust in reclaimed water, and gradually change the current incorrect impression of water reuse from the public

6.2. Develop relevant policies
Recycling reclaimed water for landscape use is part of public service, appropriate policy support from the government is necessary. Especially in the early stages of the implementation of the water reuse in the landscape, subsidies and preferential treatment should be given to stakeholders who invest and operate water reclamation facilities. Adoption of advanced and economical reclaimed water reuse technology and equipment is ought to be encouraged, so that both producers and consumers are able to benefit from the water reuse. The government and its policies should guide the reclaimed water reuse in urban landscape with the development of water reuse technology, and gradually achieve a sustainable landscape water system.

6.3. Increase the relevant design for water purification
At present, most of the landscape water design in China cities are intended towards the landscaping and improvement of the landscape yet ignore the landscape water quality maintenance and purification. With the growing shortage of water resources in the world, the reuse of water for urban landscape is bound to become a trend. Thus, while landscape designers are trying to beautify the water bodies, consideration of water quality and purification facilities are also necessary. Continuous filtration and maintenance of the recycled water would contribute to a better environment for city residents as well as a beautiful landscape.

7. Conclusion
With the increasing demand for urban landscape, landscape construction water consumption raises, and water quality standards are more stringent. Using clean water to create urban landscape not only causes a lot of waste of water resources, but also impedes the sustainable development of society.
Reasonable use of reclaimed water resources, reusing recycle water that is purification processed and meets the landscape water quality standards in landscape construction, would not only be the future trend of urban landscape development, but also able to alleviate the shortage of water resources crisis, and produce positive social effect on sustainable development. Therefore, China should pay attention to the problem of water reuse landscape, learn from the relevant foreign success stories, and develop a more effective method of water treatment. Meanwhile, exploring how to integrate water purification facilities and landscape design reasonably will become the critical issue of reclaimed water reuse in urban landscape design.

8. References
[1] WU Shuang, CHEN Yu, LUO Qiu-rolling.Current Situation of Utilization of Chinese Water Landscape and Its Enlightenment to Our Country.Value Engineering. 2014; 33 (28): 79-80.
[2] ZHANG Lin-mei, HU Cai-xia, DU Hong, ZHANG Wei-hua. Current Status and Development Prospects of Reclaimed Water. Water Science and Engineering. 2008; 2008 (1): 1-3.
[3] Xia Chengxiang. Practice and Implications of Water Recycling Abroad. China Water Resources. 2007; (7): 45-48.
[4] Wu Meng.Research on self-circulation and utilization of moat landscape water in Xi'an [Internet]. Xi'an: Xi'an University of Technology; 2008. Available from: http://cdmd.cnki.com.cn/Article/CDMD-10700-2008160327.htm
[5] Wu Meng.Research on self-circulation and utilization of moat landscape water in Xi'an [Internet]. Xi'an: Xi'an University of Technology; 2008. Available from: http://cdmd.cnki.com.cn/Article/CDMD-10700-2008160327.htm

The Distinct Character of a Capital? From Formality to Regularity of Our Ancient Capitals

T Kelly[1]

[1] Tamara Kelly, Abu Dhabi University, Abu Dhabi, UAE
E-mail: tamara.kelly@adu.ac.ae

Abstract. Many cities were shaped by dynamics political and commercial factors and man-made layers whereas other capitals have been modeled spontaneously by natural influence, London is great sample of this approach and was described as a natural city planned over decades by many hands with great appreciation to natural terrain. Contrary mode is obvious in China where several factors behind the distinct identity of Chinese architecture and urban planning. Among those influences are the Metaphysics Philosophies or Emperor guidance who encouraged uniformity in many aspects in China including city planning. The aim of this paper is to highlight the impact of various forces and mankind dogmas in shaping up a unique character of famous capitals. Hence London and Beijing are the two contradictory case studies subject to deep analysis in parallel with other theories such as Yin - Yan and Fengs Shui principles, to examine their impacts on urban planning in China.

1. Introduction

To understand the Chinese architecture the culture of China must be studied, in particular the three primary theories, QI energy, Emperor Status, Feng Shui and the five elements. Feng Shui is one of the five main Arts streams of China, meaning how to harmonize a building or any settlements with surrounding environment, to enhance flow of energy *Qi* coming from natural terrain *"The Chinese art of siting has been applied in and mankind is its guardian a very wide range of spatial dimensions, from smallest of space- say bedroom or even the location of chair to larges; cosmic direction" The History of Cartography, p216.* The building and settlements preferably positioned towards water body and at the mountain foot or towards certain star to retrieve luck. Moreover, the Chinese believe that the energy *Qi* exists in living creatures and inanimate objects including natural elements such as mountains hence most of Chinese cities allocated nearby natural feature. Figures 1 and 2. *"Mountains and rivers played an important role in the popular creeds of Chinese form the most remote antiquity as objects of worship. Mountains, forests, rivers, valleys, high hills have the power to produce rain and wind. Of all these things, it is said that they are then sacred powers," Magic Square, p 31.* Chang'an and Luoyang are major ancient capitals and where established between two natural barriers, beside the intangible benefits of natural feature, mountains are great natural defensive barrier against any invasion and a great source of construction materials similarly the water body provides water for farming and fulfil human needs in addition to acting as transportation tool to deliver labours or goods. Before laying down the settlement, the Emperor requests his officers to investigate the site and collect information regarding the availability of natural resources, soil and water quality nearby.

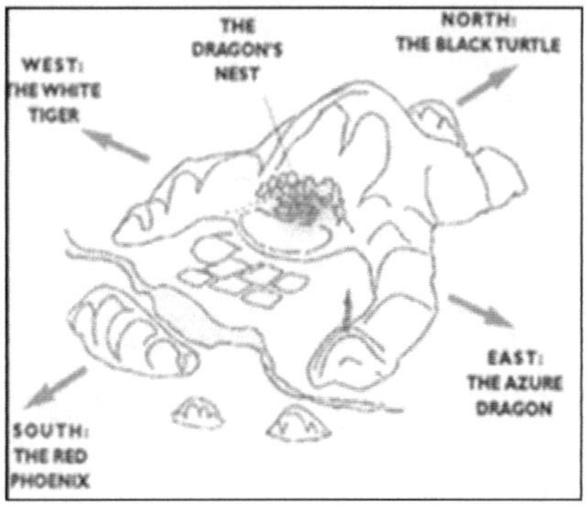

Figure 1 and 2. Settlements preferably positioned towards water body and at the mountain foot.

Qi theory is a second crucial methodology and contributor in shaping up the Chinese architecture, it has great influence on occupants' mood and health Qi is a creation of two opposite forces known as male Yan and female Yin. Those opposite forces epitomized by numbers, colors, seasons, direction temperature or natural feature figure 3. These contradicting forces form everything in our universe and should be balanced otherwise disaster and chaos will occur.

Female, Yin	Even numbers	Winter	moon	old	water	Poor, weak	distraction	Earth or valley	Black, dark	north
Male ,Y an	Odd numbers	Summer	sun	young	Fire	Strong, rich	Constrictive	Heaven, mountain	White	south

Figure 3. Yin and Yan.

Yin and Yan intermingle together to create the key elements known as Metal, Wood, Fire and Water, all living creatures or lifeless things in our universe are amalgam of those five factors. In Chinese mythologies the male force and female were created since accouchement of the cosmos and their harmonized admixture occurred at the center of our planet earths leading to initiate various deities and commenced life on earth. The fact the effect of Yin and Yan theory is tangible in building orientation, premises are usually designed to faces south to enhance air circulation and maximize heat and light attainment furthermore south is Yan the male power and a symbol of youth, wealth, and luck this lad to grid layout of city to allow south positioning of premises *"The philosophy foundation of development of square –shaped cities in ancient china was determined by ancient philosophy such as the philosophy of YING –YANG along with the principle of the five elements of water, fire, earth, wood and metal. The theme of duality, which featured in these philosophies led to an emphasize on forming a central axis in the basic layout of cities and also promoted symmetry." Chinese Architecture, p 9.*

2. Central axis and well field graph

As early mentioned the interaction of two forces at centre of earth resulted in creation of first mankind - the Emperor, who was regarded the guard of earth, in charge of keeping male and female forces in harmony. This myth promoted centrality and enormously influenced the arrangement of building in city planning, as key buildings including government buildings or Emperor's palaces were commonly located at the core of the city underpinning central Axis. Forbidden city is obvious sample echoes this principle, it was established at the core of Beijing accommodating emperor places and other administrative premises which were laid along central alignment including main gates . "*The garden was the good Earth, and mankind is its guard for this he had to establish harmony in himself between yan and yin*" Magic square, p35. Order is crucial to rule China the vast nation and the well field disarm is the upshots of central rhymester approach, in actual fact the amalgam of Chinese philosophies including, Feng shui, the five elements, Centrality, Order and Yin Yan have spawned the well field system. The diagram is a mixture of numerology, cosmology, geomancy or astrology and the blend of such notions produced the3X3 in one square graph locating the sentry the emperor at the focus of the diagram to safeguard earth. The well field diagram is the base for most of ancient cities throughout Zhou Dynasty, during this era the emperor authority overshadowed religious power hence the diagram a knowledge the setting of the Emperor at the centre between heaven and earth. The diagram accommodates numbers underlining Yin and Yan forces, the central square marked by the odd male digit 5 underscoring the five primary elements and comprises the imperial zones, ancestors and land temples exist in square 7 and 3 while the market in square 9 figure 4.

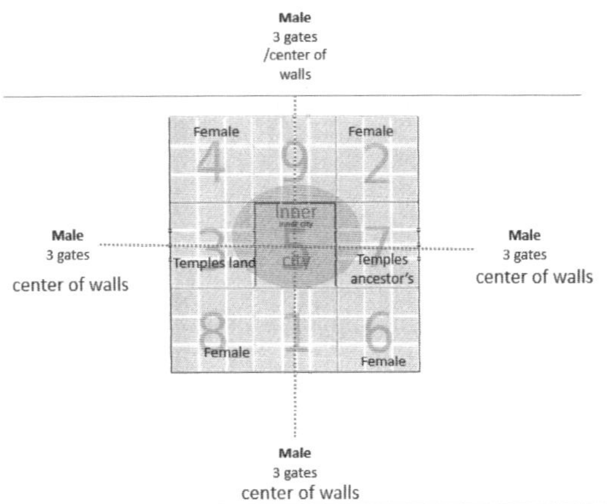

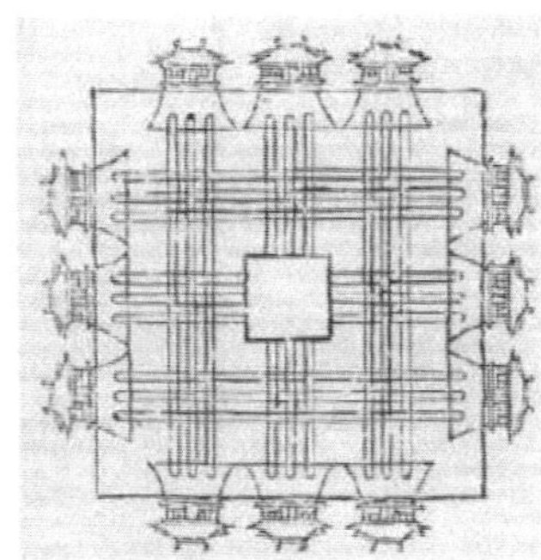

Figure 4 and 5. Well field diagram reflects Yan force.

The well field diagram underlines orderly grid and the 9 main roads run horizontally and vertically in perpendicular arrangements and first application of the diagram was marked in Cheng Zhou 1036 BC figure 5. *"This plan, like its counterparts in several other Eurasian civilization, was designed to illustrate and establish the centrality of the ruler, the son of Heaven who sat at the pivot of the four quarters and meditate between heaven and earth. Hence Emperors throughout Chinese history often concerned themselves with the geometric layout and numerical symbolism of the capital and its cultic sites and structure." The History of Cartography, P212.*

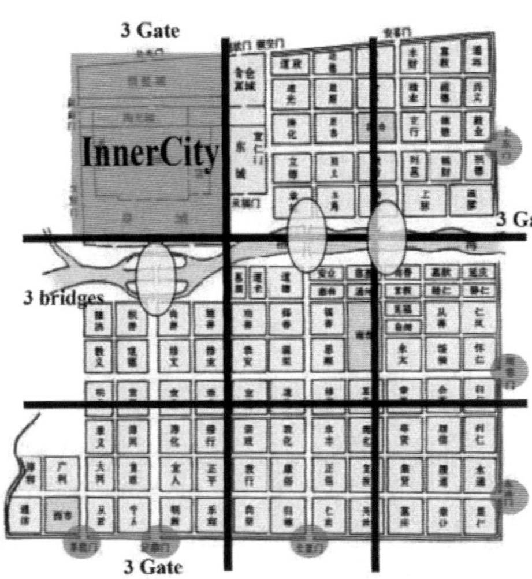

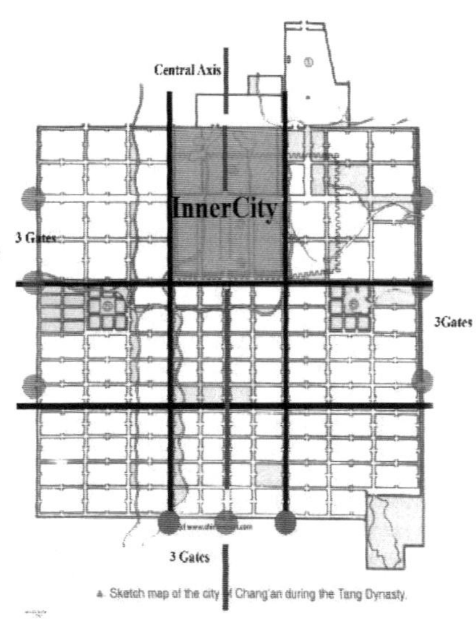

Figure 6 and 7. Luoyang and Chang'an cities.

3. Case studies

The layout of Beijing, Chang'an and Luoyang cities reveals the impact of wholly Chinese metaphysics theories. All those cities acknowledge Feng Shui theory and were established at the end of mountains and nearby rivers. The last two case studies abide by well field diagram with cites adjacent to Wei River and Yellow River but the inner city location slightly changed upwards in Luoyang.

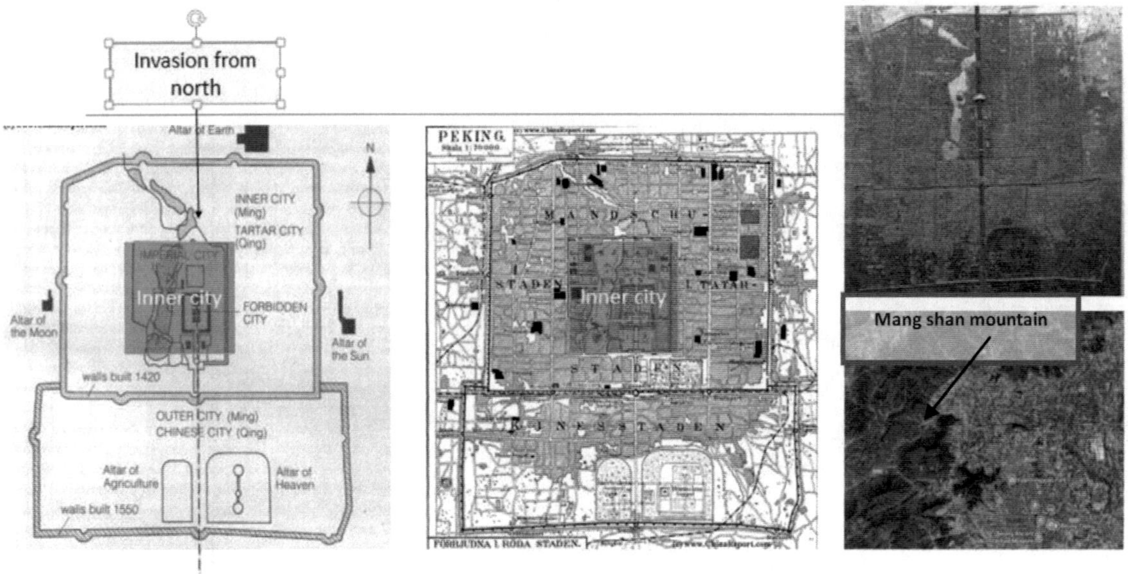

Figure 8. Beijing city, the imperial zone undely the centraal axis.

The logic of such alteration is clear as inner city is armoured by two natural barricades, the mountain Mangshan north east embraces imperial zone and is a great barrier while Luo River is southwards segregating internal city form outer district or any invasion coming from South. Furthermore, the river provides water supply for drinking or farming and Yan force underscored by 3 crossing bridges over the river figure 6. The wall surrounding the city accommodates three gates in each side apart from east side and similar arrangement is perceptible in Chang'an figure 7 in Chang'an city same rational applicable as Qinling mountain is a natural shielding border extending north protecting inner city from possible conquest coming across the north whereas urban fabric beneath inner city is a further obstruction if invasion occurs from southern direction. Beijing city is a third case study adhered the well field layout, Forbidden City occupies the centre and the influence of Yan forces is apparent most of the buildings within the imperial zone grouped in odd numbers 3 or five "*The compound is divided into two complexes along the North – South axis; the southern outer court has three principal halls while the northern inner courtyard has another three main buildings" Chinese Architecture, p3"*. South alignment is acknowledged as main entry gates of palaces are through south- north axis additionally temple of Heaven located southern Forbidden City and the journey to it is through the southern axis. Figure 8.

4. London the sporadic city

The great city of London is the opposite of Chinese orderly grid layout cities, London is a natural capital shaped over the eons by many layers with excessive appreciation to its natural features including; natural topography, layers of green infrastructure of hills, tidal flows, Thames River and its tributaries. The Thames exists since centuries and it is a key component of London's urban fabric and crucial contributor in determining the form of settlements around its banks. In fact the river can be divided into three crucial segments:

- The core part is dense urban where government buildings and historical landmarks are situated
- The western part is Rural and famous for green parks

- The eastern part as a linking district with Europe through the sea. Eastern Thames accommodated harbours and great industrialized settlements built in 19th century as point of trade off zones with the world beyond the sea.

The logic behind such setting is the nature of river, the north bank along the middle part of Thames is higher than south and this meant flood risk is less, more sunlight and warmth gain in adjacent building in north bank. Hence the land here is expensive and government building or many historic landmarks of London have been erected in this share figure 9. Romans during their invasion of London recognized that the distance between the two banks is lest at the centre of Thames thus they composed the first crossing bridge in Lambeth and established their early settlements at middle Thames. In fact those are the forces and reasons why London was originated in this part and grew for decades within the bents of Thames from Westminster to pool of London figure 10.

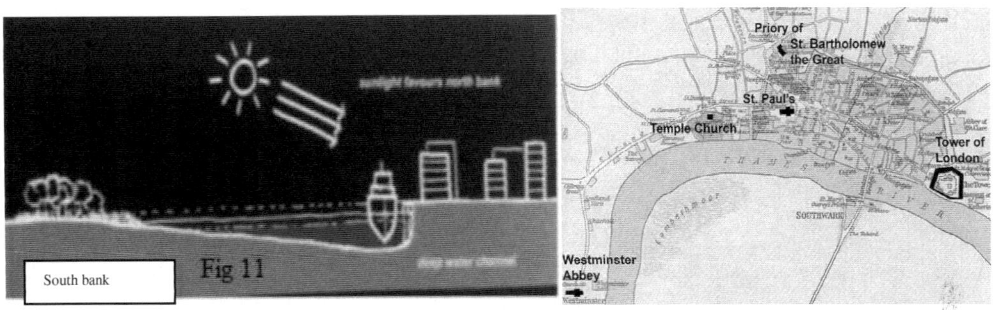

Figure 9 and 10. Thames's banks and left figure London City during Romans invasion.

In comparison Southern bank of Thames is cheaper flusher with lower elevation and commonly sludge with other materials mount upon the edge and the access into the bank strict without lath. That effected the value of the land thus was uninhabited for long period. It was subject to several flooding events. It urbanized slowly in comparison with north bank figure 11.

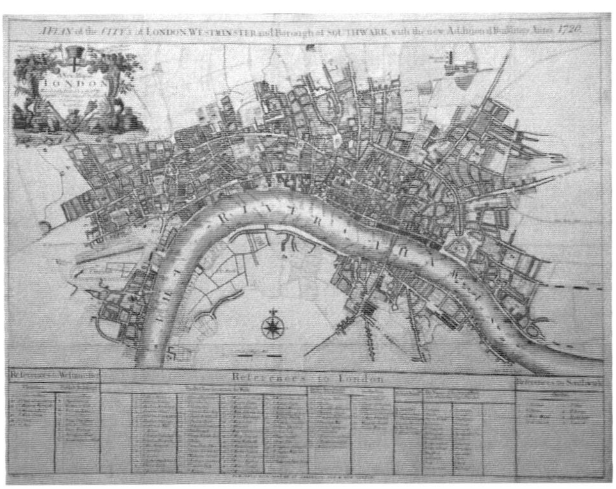

Figure 11. Development of London 1780.

In 1661Vauxhall garden opened a leisure garden that accommodated concert orchestra side shows beside masquerades. Then gradually southern side of Thames accommodated pleasure and entertainment components such as parks, globe and rose theatres, bull baiting, prostitution activities, warehouses, drinking and gambling. *"The largest cultural complex in Europe, the festival Hall and National theatre that has been active since Shakespeare's time. On the south bank, South of the Thames River are also the cherry gardens" shaping London p37.* Later during the industrial revolution the south bank identity changed; Vauxhall garden closed and different building types added including warehouses, wharfs, breweries, and private owned plots. Later in 1917, south bank quality was altered by creating a public frontage and river walkways to enhance its image and nowadays the river still accommodates great celebration events as new-year fireworks. The south bank is famous for the Mayor's Thames Festival in September which is regarded the biggest public festival in London. *"The south bank today has moved from being a world of introverted, inaccessible private utility to its opposite, a world of public recreation with access to all"* Shaping London, p54. East Thames is inaccessible owned by companies and recognized as a connection ward with the world hence quays, offloading dock facilities or warehouses were dominant buildings types in this part. Immense trading activities evolved and numerous wharves constructed over decades to deal with goods coming from various countries such as Canada, Russia and Scandinavian countries, figure 12. Consequently an industrial district existed around the ports accommodating cranes, locks and gates, edges, water proof ponds and security high walls. Over the time Inwards looking Communities created around the river's bent opposite the mud flats accommodating cheap housing, shops, pubs, cafes to serve the workers and sailors, in fact this vast complex of dockland was great phenomena in its time on many levels, figure 13.

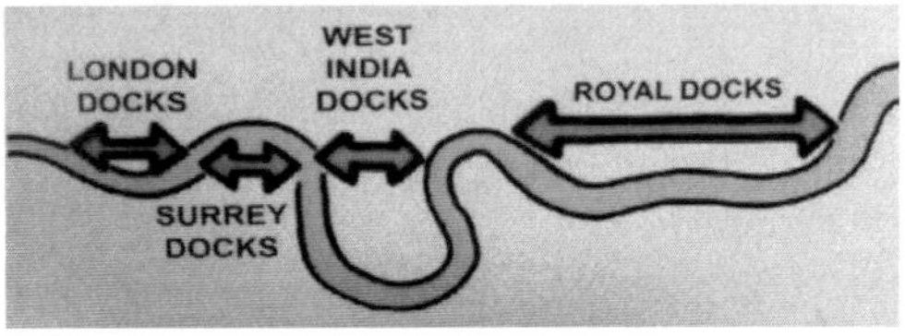

Figure 12. Docks in estren thames.

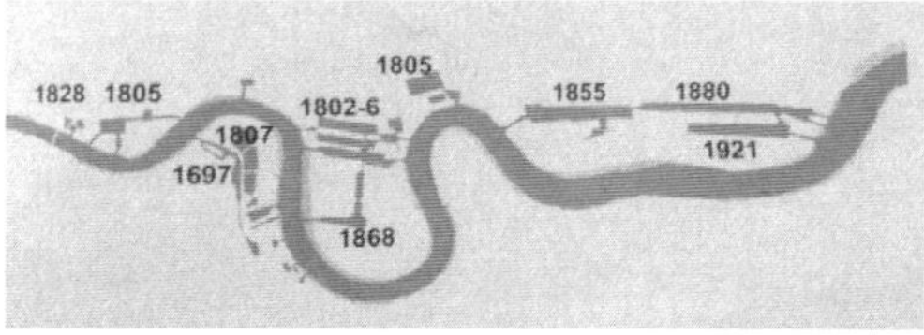

Figure 13. Docks construction.

Lastly, western Thames is distant from poluted estren comercial industrial hub and opens towarde countryside as quality of life is better unpoluted remote from congested centre of London hence and up to nowadays western Thames is a pleasant expensive area. "*Richmond palace and Hampton court were within the easy river transportation from the palace of Westminster, but away from prevailing wind that blew pollution and smoke downstream eastwards and out to sea" shaping London p26.* Moreover the River with its secondary streams provide needed water to set up and mantain many parks located on the low banks on inside bent of the river, whereas the settlements are on the outer higher side of the bends. The Thames in western part in comparison with central urban bank is supplementary accessible thru several parks including Hampton Park, Bushy, Richmond and Marble hill park, old deer park, Kew garden and botanical garden. Those parks are vital municipal landscape woven within the bends of the river providing vital accesses to water body and endowing openness platforms to link with nature figure 14.

The tidal flow and flood risk is limited in north west bank and the availability of vast open public parks in Western Thames fortified wealthy citizens to reside western part hence the urban fabric in western Thames varieties from luxurious palaces to smaller nobility villas and houses or bungalows moreover houseboats are remarkable elements of Western Thames. Accordingly this assortment of premises and types of settlements in western part contribute in shaping distinct multi layered character of Thames and diversify scenery of upper Thames. "*Bend in the river, the alternating of landscape and urbanization, the frequency of islands breaking the stretches of water into reaches of different scale all of this establish a pattern, a rhythm that makes this glorious section of metropolitan Thames area " shaping London*".

4.1 The impact of water body in shaping the identity of London

There are too many secondary streams linked to the Thames and those branches of water were key elements in forming London city, the 55 tributaries cultivating local communities and agricultural settlements originated around them till 18th as they provide water supply for farming and drinking, sewage dumping and the power of water transformed by mills. Then after industrial revolution the land use was altered to accommodate industrial and residential developments for middle class workers and the usage of water evolved to create numerous remarkable parks where the streams regarded a feature of many parks. Westbourne River is a major componenet of Hyde park, additionaly, Lea River in East London runs through victoria park and Tyburn is a key element in Regent's park whearase Revensbourne and beveleybrook streams are crucial in many parklands in the South, including Richmond and bushy parks figure 14. The urban grain between early mentioned four rivers differ according to layout and direction of rivers, hence developments appeared as piece of geometrical fabric located between irregular water courses figure 15 .

In fact London unlike many European cities that were shaped by mankind hand from scratch, following formal grid arrangement, the layout of London is extremely shaped by land form and natural bent tributaries channelled towards the Thames, thus the water courses are first layer and the land in between formed accordingly over decades. *"the genuius loci of so much of central London.The streams beds are the first layer and the field in between are the second, which were laid out as residential estates and the crown lands of royal parks-StJames park, Hyde park/ kinsnigton Gardens and Regent's Ppark furher to North. This makes up the essence, the geopmatry of what we know of the core ,the heart of London " Shaping London, p107".* Rivers of London are influential and they determine the uses of the land in between Lea River East London was a crucial route trading and transporting things worldwide throughout industrial era hence specific developments established in the land between waterbody. Whilst Thames Width and depth were main elements effecting the location of first part of London City, furthermore Thames shaped the layout of London thoroughfares as main arteries grows outwards in parallel ripple arrangement from the River Akin to annular encircle of trees, for instance Whitehall and Stand Streets are the first adjacent parallel ring to Thames accommodating ceremonial events. Thru early mentioned discussion the water body resulted in such natural layout of London the capital and Rivers were key elements affecting land uses, urban grain and resident's stratum.

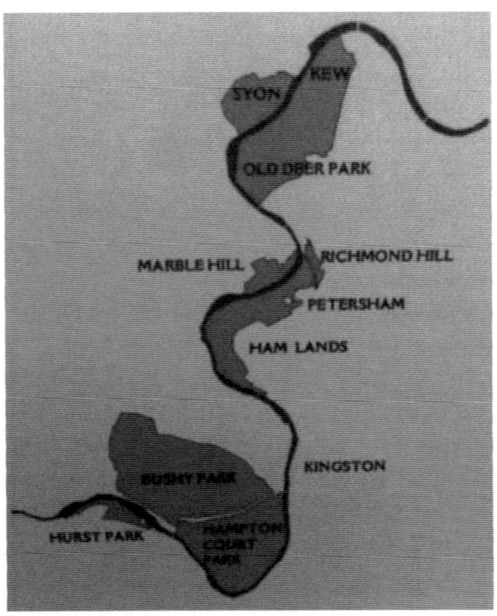

Figure 14. Parks along Western Thames.

Figure 15. River Thames and its streams.

5. Conclusion

The Thames and its tributaries were and still a great contributor in shaping the layout of London city, water channels were vital Catalyst for transportation and farming. Romans initiate London City near an area where the river is deep enough for vessel movements and narrow to connect by bridges thus London as a city grew up around the Halbrook area on the north bank as the land and sun light gain is high consequently flood risk is less. The land was precious inhabited by nobles and places, historical landmarks or vital government buildings established in northern part. West of Thames is distant from eastern commercial industrial zones and dense congested central urban fabric moreover West London accommodates openness landscape and availability of vast green parks with good quality environment raised the value of land. The negative impact of natural terrain is visible in southern bank where the land is low and subject to several flood events this effected urban fabric and the area was developed slowly in comparison with northern bank and land uses were limited to entertainment activities thus watercourses and natural geographies were first layers then manmade urban fabric woven between irregular lines of streams forming a distinct natural Capitals. London is the opposite approach of Ancient Chinese capitals when the impact of metaphysic theories and Emperor Status led to emphasizing central axis and orderly grid layout cities evolved during Zhou Dynasty and influenced the outline of many Chinese capitals during different dynasties. Most ancient capitals reflect Yin and Yan theory, Feng Shui and the five primary elements principle this resulted in locating settlement nearby water body and natural terrain to utilize Qi energy. While acknowledgment of Yin Yan forces led to south facing preferences and odd numbers were dominant feature representing male power and commonly used in grouping building.

6. References

[1] Cai Yanxin. Chinsese Architecture, Cambridge university press (2010) 9-47
[2] Michael Frenzo,Marian Moffett, Lawrance Wodehouse. World History of Architecture, Laurence king publisher (2013) 81-91.
[3] J.B.Harley and D.Woodward. The History of Crtography, University of Chicago press (2010) 203 2018.
[4] A er Lei De Sheng . Magic squares: the ancient Chinese city.(1991) p5-50.
[5] Geremie R. Barme. The Forbidden City -Wonders of the World (2012)p 10-70.
[6] Simon Foxell, Mapping London, Dlock Dog Publishing 2007
[7] Stephen inwood, A history of London, Peter Macmillan, 1998
[8] Ackroy, Thames Sacred River, Chatto &Windus, 2007,
[9] JOH Bunge. Tideless Thames in future London, the Thames barrage association, 1997
[10] Gavin Weightman, London River: the Thames story, Collins &brown, 1990
[11] Terry Farrell, Shaping London, Willy, 2010

An Evaluation Model for Sustainable Development of China's Textile Industry: An Empirical Study

Hong ZHAO[1], Xiaodong LU[2], Ting YU[3] and Yanbin YIN[4]

[1] Professor, School of Economics of Tianjin Polytechnic University, Tianjin, China
[2] Lector, School of Textiles of Tianjin Polytechnic University, Tianjin, China
[3] Student, School of Economics of Tianjin Polytechnic University, Tianjin, China
[4] Professor, School of Economics of Tianjin Polytechnic University, Tianjin, China
E-mail: zhaohong300160@126.com

Abstract. With economy's continuous rapid growth, textile industry is required to search for new rules and adjust strategies in order to optimize industrial structure and rationalize social spending. The sustainable development of China's textile industry is a comprehensive research subject. This study analyzed the status of China's textile industry and constructed the evaluation model based on the economical, ecologic, and social benefits. Analytic Hierarchy Process (AHP) and Data Envelopment Analysis (DEA) were used for an empirical study of textile industry. The result of evaluation model suggested that the status of the textile industry has become the major problems in the sustainable development of China's textile industry. It's nearly impossible to integrate into the global economy if no measures are taken. The enterprises concerned with the textile industry status should be reformed in terms of product design, raw material selection, technological reform, technological progress, and management, in accordance with the ideas and requirements of sustainable development. The results of this study are benefit for 1) discover the main elements restricting the industry's sustainable development; 2) seek for corresponding solutions for policy formulation and implementation of textile industry; 3) provide references for enterprises' development transformation in strategic deployment, fund allocation, and personnel assignment.

1. Introduction

With a continuously rapid growth in China's economy, it is required to search for new rules, summarize experiences, and adjust strategies for the textile industry. Sustainable development of the textile industry is a comprehensive research subject, as it is related to dynamic balances and common coordination among different aspects of the textile industry, in terms of the economy, resources, the environment, and ecology.

In this paper, the Analytic Hierarchy Process (AHP) and Data Envelopment Analysis (DEA) approaches were used for an empirical study of China's textile industry. The economic condition, social environment, natural ecology, and scientific technology as related to sustainable development were taken into consideration [1]. The natural ecology, providing raw materials for the sustainable development of the textile industry, forms the material basis as well as the fundamental factor. Economic condition is the vigor of the industry's development. A proper social environment facilitates the supply of human resources and creates industrial demands. Scientific technology allows for innovation that drives a sustainable development. These four closely coordinated factors stimulate the textile industry's development by influencing each other.

2. Evaluation model

Analytic Hierarchy Process (AHP), based on a multi-objective structure, was used to calculate and evaluate the importance of several alternatives depending on subjective judgment. Evaluation targets included target hierarchy, criterion hierarchy and project hierarchy, in which evaluating variables were compared in a judgment matrix and a numerical matrix. As a common method in empirical analysis, AHP allows for processing and qualitative sorting of quantitative data, and the calculation of the evaluation index weight and standardized weight of the evaluated variables.

Data Envelopment Analysis (DEA), an evaluation method based on the concept of "evaluating the relative efficiency" by A. Charnes and W.W.Cooper (the -C^2R Model), first appeared in 1978 and gradually evolved into a common analysis method in management science and system engineering. DEA can measure the relative efficiency of decision making units (DMUs) of the same type. The "projection principle" can do further analysis on non-DEA effective causes of DMUs to provide references for decision makers. DEA was implemented as follows: picking up a series of input and output indexes as DMUs, analyzing effective factors by the Duality Theory, and evaluating the relevant indexes. The weight was calculated out of realistic data during this process, allowing for an objective analysis.

This paper employed the DEA/AHP approaches according to characteristics of the textile industry. Evaluation models of DEA/AHP were divided into two phases: first, picking up two DMUs in one team and calculating its efficiency value by DEA; second, building an AHP matrix by ranking all the DMUs by the single hierarchy analysis method based on the efficiency value in the first phase, and in the end transforming all the AHP ranking values into DEA/AHP efficiency values. This model emphasized the objective changes of the index weight to avoid any subjective judgment. In this model, DEA was used to improve the calculation efficiency of AHP by putting every efficiency value as decision-making threshold value, and the single hierarchy analysis method in total ranking of the AHP judgment matrix contained several multiple-level hierarchical judgment and decision making efficiency values, in order to guarantee an objective reference value. The model was built with MATLAB, as follows:

2.1 Building judgment matrix

[2] Given: there are n DMUs and m input indexes (negative indexes) in each DMU, and s output indexes (positive indexes). Given: *Xij* is the jth input index of the ith DMU. Pick up two DMUs A and B and confirm the DMU relative efficiency through cross-efficiency evaluation. E_{AA} is the objective function's optimal solution of LP_1 and E_{AB} is of LP_2. The model can be described by these equations:

$$LP_2, st\begin{cases} \sum_{i=1}^{m} v_i x_{iB} = 1 \\ \sum_{r=1}^{s} u_r y_{rB} \le 1 \\ \sum_{r=1}^{s} u_r y_{rA} - E_{AA} \sum_{i=1}^{m} v_i x_{iA} \le 0 \\ u_r \ge 0, r = 1, 2, \dots, s \\ v_i \ge 0, i = 1, 2, \dots, m \end{cases} \tag{1}$$

$$LP_1, st\begin{cases} \sum_{i=1}^{m} v_i x_{iA} = 1 \\ \sum_{r=1}^{s} u_r y_{rA} \le 1 \\ \sum_{r=1}^{s} u_r y_{rB} - \sum_{i=1}^{m} v_i x_{iB} \le 0 \\ u_r \ge 0, r = 1, 2, \dots, s \\ v_i \ge 0, i = 1, 2, \dots, m \end{cases} \tag{2}$$

$$E_{AA} = \max \sum_{r=1}^{s} u_r y_{rA} \tag{3}$$

$$E_{BA} = \max \sum_{r=1}^{s} u_r y_{rB} \tag{4}$$

where u_r is input index weight and v_i is output index weight; E_{AA} is efficiency value of A and E_{BA} is cross-efficiency value of B. According to the DEA/AHP model, $E_{AA} = E_{AB}$, $E_{BA} = E_{BB}$.
The AHP judgment matrix was built based on the model. a^{jk} is the AHP judgment matrix and $C=[a_{jk}]_{n*n}$ is the element in Line I and Row K.

$$a_{jk} = \frac{E_{jj} + E_{jk}}{E_{kk} + E_{kj}} \tag{5}$$

$$a_{jk} = a_{kk} = 1, a_{kj} = \frac{1}{a_{jk}}$$

$A=|a_{jk}|_{n*n}$ is the result of effective analysis of every DMU in the DEA model.

2.2 Total ranking in the AHP model
Total ranking of DUMs through a single hierarchy process in the AHP judgment matrix was performed. As for $A_{n*n}=[a_{jk}]_{n*n}$, λ_{max} and w:$w=(w_1,w_2,\ldots, w_n)^T$, they were calculated by the root method, referring to the relevant importance of the j_{th} DMU. The m_{th} component w_m of the eigenvector w reflects the relevant importance of the m_{th} DMU, which is the total ranking value in the AHP model. The total ranking value in AHP was turned into a DEA/AHP efficiency value. That is, the biggest component of the eigenvector w was chosen and given an efficiency value of l, and the corresponding efficiency values, dividing l by other components, were obtained. The details are shown in equation (6):

$$a_{jk} = \frac{E_{jj} + E_{jk}}{E_{kk} + E_{kj}} \tag{6}$$

$$a_{jk} = a_{kk} = 1, a_{kj} = \frac{1}{a_{jk}}$$

For AHP, first, some values were set：
The weight factor of n factors was W：

$$W^T = \begin{bmatrix} w_1 & w_2 & \ldots & w_n \end{bmatrix} \tag{7}$$

The relevant importance matrix A was obtained out of comparing every two factors of n： $A=(a_{ij})$
Among： $a_{ij}=w_i/w_j$ $a_{ji}=1/a_{ij}$
Then： If A is a consistency matrix, n is its eigenvalue and W is its eigenvector. Hence,
$AW=\lambda W$ $\lambda_{mac}\geq n$

2.3 Model application
The model can be applied by comparing some evaluation indexes affecting textile industry, such as total industrial discharges, waste water facilities and industrial waste gas emission. The comparison can be done as follows:

$$A = \begin{bmatrix} 1 & 5 & 3 \\ 1/5 & 1 & 1/3 \\ 1/3 & 3 & 1 \end{bmatrix}$$

Summation

1) Keep the total of every row 1.

$$b_{ij}=\frac{a_{ij}}{\sum_{i=1}^{n}a_{ij}} \qquad B=\begin{bmatrix}0.652 & 0.556 & 0.692\\ 0.130 & 0.111 & 0.077\\ 0.218 & 0.333 & 0.231\end{bmatrix}$$

2) Keep the total of every line 1.

$$v_i=\sum_{j=1}^{3}b_{ij} \qquad V=\begin{bmatrix}1.900\\ 0.318\\ 0.782\end{bmatrix}$$

3) Normalization method.

$$w_i=\frac{v_i}{\sum_{i=1}^{3}v_i} \qquad W=\begin{bmatrix}0.633\\ 0.106\\ 0.261\end{bmatrix}$$

$$AW=\begin{bmatrix}1 & 5 & 3\\ 1/5 & 1 & 1/3\\ 1/3 & 3 & 1\end{bmatrix}\begin{bmatrix}0.633\\ 0.106\\ 0.261\end{bmatrix}=\begin{bmatrix}1.946\\ 0.319\\ 0.787\end{bmatrix}$$

4) Consistency examination.

$$\lambda_{\max}=\frac{1}{n}\sum_i\frac{(AW)_i}{w_i} \qquad C.I.=\frac{\lambda_{\max}-n}{n-1}$$

$$\lambda_{\max}=\frac{1}{3}\left[\frac{1.946}{0.633}+\frac{0.319}{0.106}+\frac{0.787}{0.261}\right]=3.033$$

$$C.I.=\frac{3.033-3}{3-1}=0.0165 \qquad C.R.=\frac{C.I.}{R.I.}$$

R. I. is average random consistency index,$C^2R<0.1$ means the acceptance of the consistency of the judgment matrix.

5) Synthetical importance: turn the relevant weight into evaluating point by normalization method.

$$S_{ij}^{k}=\frac{W_{ij}^{k}}{\max\left|W_{ij}^{k}\right|}\beta$$

where S^k_{ij} is the evaluating point value of the cross point of Line I and Row J in Hierarchy K, and β is the largest evaluating point value.

Note that in the above, S*10=[10 1.67 4.12]

2.4 Model analysis

DMU is the manufacturing process in which a set number of productive factors are input to produce the largest quantity of products (the ith DMU is usually remarked as DUMi). Hence DMU refers to a process turning input into output.

With the C^2R Model in DEA, one can first do an analysis on the input and output indexes of the chosen DMUs. A DMU takes in many inputs and outputs. In table 1, there are n DMUs of the same type and every DMU has m kinds of input factors and s kinds of output factors. The input and output vectors of DMU are shown in table 1.

Table 1. Table of input and output vectors of DMU

	No.	Weight	DMU					
			1	2	…	j	…	n
Input data	1	v_1	x_{11}	x_{12}		x_{1j}	…	x_{1n}
	2	v_2	x_{21}	x_{22}		x_{2j}	…	x_{2n}
	…	…	…	…	…	…	…	…
	m	v_m	x_{m1}	x_{m2}		x_{mj}	…	x_{mn}
Output data	1	u_1	y_{11}	y_{12}		y_{1j}	…	y_{1n}
	2	u_2	y_{21}	y_{22}		y_{2j}	…	y_{2n}
	…	…	…	…	…	…	…	…
	s	u_m	y_{s1}	y_{s2}		y_{sj}	…	y_{sn}

x_{ij} is the data of the ith input in the j_{th} DMU,

y_{ij} is the data of the i_{th} output in the j_{th} DMU, v_i is the weight of the i_{th} input,

u_y is the weight of the i_{th} output, i=1,2,…,m r=1,2,…,s j=1,2,…,n

Given $X_j=(x_{1j},x_{2j},\ldots,x_{mj})^T$, $Y_j=(y_{1j},y_{2j},\ldots,y_{sj})^T$ $V=(v_1,v_2,\ldots,v_m)^T$, $U=(u_1,u_2,\ldots,u_m)^T$

The efficiency evaluation index equation of the Jth DMU is:

$$E_j = \frac{U^T Y_j}{V^T X_j}$$

By performing efficiency evaluation of the $j_{0\,th}$ DMU, the C^2R model is:

$$\begin{cases} \max \dfrac{U^T Y_0}{V^T X_0} \\ \dfrac{U^T Y_j}{V^T X_j} \le 1 \\ u \ge 0 \\ v \ge 0 \end{cases}$$

In order to judge the validity of DMU, dual problem of linear programming can be used, and the dual model of C^2R model becomes:

$$\begin{cases} \min \theta = \theta^* \\ st \sum_{j=1}^{n} \lambda_j x_j + s = \theta x_0 \\ \sum_{j=1}^{n} \lambda_j Y_j - s = Y_0 \\ s^- \ge 0, s^+ \ge 0 \end{cases}$$

The basic variables of C^2R model are listed as follows:
$\theta$is efficiency value of DMU, X_j:a set of input factors of ; and $X_j = (x_j, x_{2j}, \ldots, x_{mj})$.$Y_j$: a set of output factors of DMU, and $Y_j = (y_j, y_{2j}, \ldots, y_{sj})$; λ_j is the portfolio ratio of the j_{th} DMU in the rebuilt DMU combination DMU_0; $S^-$and S^+ are slack variables.
The optimal solutions of the dual program includeλ^*,θ^*, $S^{*-}$and S^{*+}, in which$\theta^*$is the relevant efficiency value of DMU. The economic implications of the C^2R model refer to the following:

- if the evaluated DMU is relevantly efficient, only when$\theta^*=1$ and all the optimal slack variables $S^{*-}= S^{*+}=0$, the output Y_0 of the economic system constituted by n DMUs based on the initial input X_0 can get to the optimal value.
- if$\theta^*=1$ and$\theta^*\neq 0$ or $S^{*+}\neq 0$, DMU is weakly efficient in DEA. In the economic system constituted by n DMUs, the initial output Y_0 can be kept invariant when the input X_0 decreases S^-; the output can increase S^+ when the input X_0 remains invariant.
- if $\theta^*<1$, DMU is inefficient in DEA: In the economic system constituted by n DMUs, the initial output can be kept invariant with the input reduced to theθ^*% of the initial input.

3. Construction of the evaluation index system

This empirical study focused on how the source input of environmental protection in the textile industry (in this paper we primarily referred to the clean energy production) could affect wastewater reduction and the reduction's influence on the industry's international competitiveness. The representative indexes were analyzed, which reflected implementation of clean energy production investment, source treat capabilities and China's textile industry's international competitiveness at this stage [3].
A common index system can be established in the E-R model, in which frequency statistics, theory analysis and setting indexes developed by consulting experts according to scientific and reasonable principles. In order to show the operability and flexibility of the evaluation index system, regional textile industry's development needed to be taken into consideration to evaluate the availability and reliability of the indexes under the influence of society, economy and environment. The evaluation system is illustrated in figure 1.

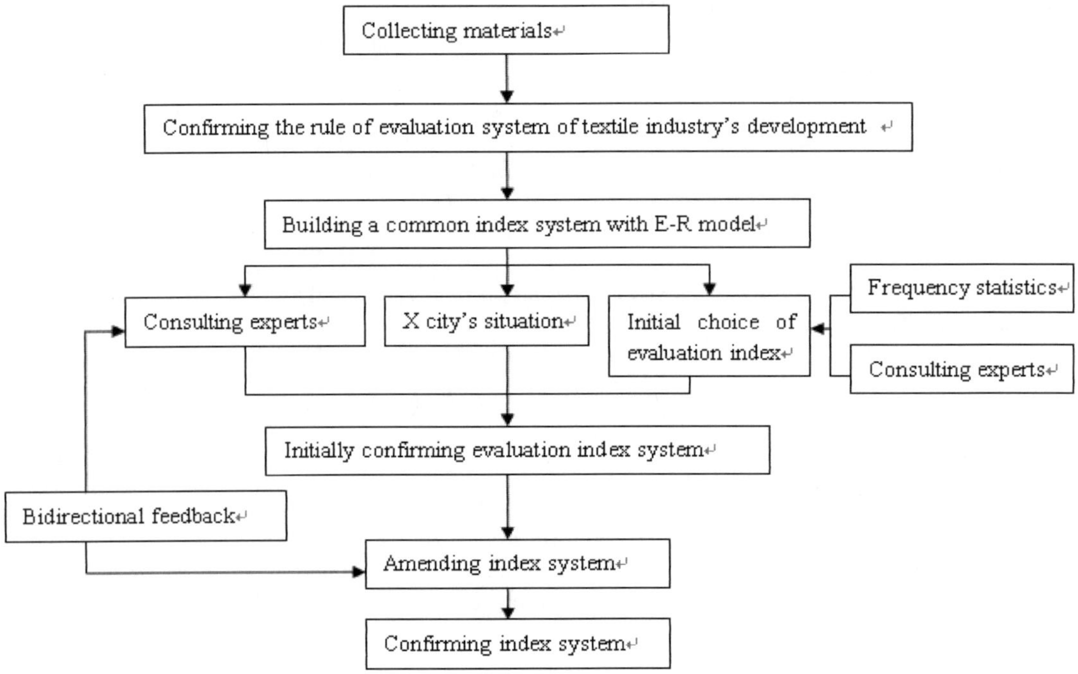

Figure 1. Flow chart of building evaluation index system of the sustainable development of X city's textile industry.

This index system consisted of 4 hierarchies. Target hierarchy A was the comprehensive index of the sustainable development of X city's textile industry, which reflected the coordination among the development situation and economy, society, and environment. Criterion hierarchy B included four parts: B_1 (index of the sustainable development of textile industry), B_2 (coordination degree between sustainable development of the textile industry and society), B_3 (coordination degree between sustainable development of the textile industry and economy) and B_4 (coordination degree between sustainable development of the textile industry and environment). Restraint hierarchy C reflected indexes restraining Hierarchy B and it covered three aspects: C_1 (the status quo of the textile industry), C_2 (development degree) and C_3 (development situation). Due to the evaluated region and evaluation focus, the evaluation methods for sustainable development of the textile industry were different in the choice of the index system.
The final, chosen evaluation indexes are shown in table 2.

Table 2. Evaluation index system of textile industry's sustainable development

1st class indexes	2nd class indexes	3rd class indexes	
Indexes of textile industry's sustainable development	Indexes of economic benefits	Indexes of gross scale	Total production value
			Used capital value
			Ratio of trade value in GDP
		Construction indexes	Ratio of finished products
			Ratio of product materials
		Efficiency indexes	Contribution rate to GDP

			Pulling degree to FDI increase
	Indexes of social benefits	Employment of textile industry	
		Average salary of textile industry	
	Indexes of ecological benefits	Net emissions of waste water of textile industry	
		Net emissions of waste gas of textile industry	
		Net discharges of solid wastes of textile industry	
		Net emissions of energy of textile industry	

4. Calculative process of evaluation model

The indexes for sustainable development of the textile industry included a large number of economic benefits indexes, social benefits indexes and ecological benefits indexes. In this study only ecological benefits indexes were being analyzed. [4] To realize the ecological benefits indexes for sustainable development of the textile industry, data needed to be analyzed from different aspects. Given three of the indexes were decisive (C_1, C_2 and C_3), some main influencing factors of textile industry were chosen, such as the total number of enterprises, total discharges of industrial waste water (ten thousand tons), standard discharges of industrial waste water (ten thousand tons), waste water treatment facilities (set), processing power of waste water treatment facilities (ten thousand tons/day), operating expense of the year (ten thousand yuan), emissions of industrial waste gas（million standard cubic meters）, fuel burning, production technology, emissions of industrial smoke (ton), emissions of industrial SO_2 (ton), emissions of industrial dust (ton), standard emissions of industrial SO_2 (ton), standard emissions of industrial smoke (ton), standard emissions of industrial dust (ton), waste gas treatment facilities (set), desulfurization facilities (set), total of industrial solid wastes (ten thousand ton), discharges of hazardous wastes and industrial solid wastes (ton), comprehensive quantity of industrial solid wastes (ten thousand tons), disposal quantity of hazardous wastes and industrial solid wastes (ten thousand tons), ratio of hazardous wastes and industrial solid wastes, comprehensive reusing value of wastes (ten thousand yuan), and so on. In AHP, priority ranking of the influencing factors for sustainable development of the textile industry can be performed (the weight and calculation procedures are given) [5].

A, B, and C respectively referred to the three hierarchies of the influencing factors. The paired comparison method was employed in the model to obtain the following judgment matrices as shown in tables 3- tables 5.

Table 3. A-B Judgment matrix

A	B_1	B_2	B_3	B_4	B_5	B_6
B_1	1	1	1	4	1	1/2
B_2	1	1	2	4	1	1/2
B_3	1	1/2	1	5	3	1/2
B_4	1/4	1/4	1/5	1	1/3	1/3
B_5	1	1	1/3	3	1	1
B_6	2	2	2	3	3	1

Table 4. B_1 B_2 B_3 - C Judgment matrix

B_1 B_2 B_3	C_1	C_2	C_3
C_1	1,1,1	1/4,1/4,3	1/2,1/5,1/3

C_2	4,4,1/3	1,1,1	3,1/2,1/7
C_3	2,5,3	1/3,2,7	1,1,1

Table 5. B_4 B_5 B_6 - C Judgment matrix

B_4 B_5 B_6	C_1	C_2	C_3
C_1	1,1,1	1/3,1,7	5,7,9
C_2	3,1,1/7	1,1,1	7,7,1
C_3	1/5,1/7,1/9	1/7,1/7,1	1,1,1

In AHP, considering the influencing factors for sustainable development of the local textile industry in China, the first factor was a restricted factor needing key treatment. Hierarchy total ranking was obtained and its result was the weight of the evaluation index system (table 6). In this table, C_1 refers to the evaluation weight of the status quo of textile industry, C_2: the evaluation weight of developing degree, and C_3: the evaluation weight of developing situation. The total ranking result of B1 to B6 indicated the weight of indexes of the industry's influencing factors.

Table 6. Hierarchy total ranking

	B_1 0.1507	B_2 0.1792	B_3 0.1886	B_4 0.0472	B_5 0.1464	B_6 0.2879	Hierarchy total ranking
C_1	0.1365	0.0974	0.2426	0.2790	0.4667	0.7986	0.3952
C_2	0.6250	0.3331	0.0879	0.6491	0.4667	0.1049	0.2996
C_3	0.2385	0.5695	0.6694	0.0719	0.0667	0.0965	0.3052

Table 6 shows that the status quo of textile industry (C_1) is the highest weight in the overall evaluation of the three indicators system, thus becoming the most important problem for China's textile industry to promote the sustainable development of economy.

5. Conclusion

In this paper, AHP and DEA approaches were used for an empirical study of China's textile industry. The economic condition, social environment, natural ecology, and scientific technology were used to establish the models. The AHP model prioritized the factors influencing the sustainable development of the textile industry. The results suggested that the status of the textile industry has become the major problems in the sustainable development of China's textile industry. The enterprises concerned with the textile industry status should be reformed in terms of product design, raw material selection, technological reform, technological progress, and management, in accordance with the ideas and requirements of sustainable development.

6. References

[1] Ma, Chuandong. Economy of Sustainable Development [M].Jinan: Shandong People Press. 2011:332-336.
[2] Zhao，Junli. Analysis of China's industrial upgrading based on the adjustment of American textile industry [J]. International Trade.2010:131-136
[3] Zhang, Fengtao. Research on Cluster Competitiveness of China's Textile Industry [D]. Northeast Normal University. 2012:31-36.

[4] Dong, Wenwen. Export Status pf China's Textile Industry and Measures for Sustainable Development [J].Foreign Trade. 2013:11-12.
[5] Ding, Jue. Research on Factor Intensity Reversal and China's Textile Industry's Upgrading [D].Zhejiang University. 2010:54-58.

IOP Conf. Series: Journal of Physics: Conf. Series **989** (2018) 012009 doi:10.1088/1742-6596/989/1/012009

Sustainable and ICT-Enabled Development in Developing Areas: An E-Heritage E-Commerce Service for Handicraft Marketing

Tufail Muhammad[1] and KwanMyung Kim[2]

[1] PhD Candidate, Graduate School of Creative Design Engineering, Ulsan National 2 Institute of Science & Technology (UNIST), Ulsan, South Korea
[2] Associate Professor, Graduate School of Creative Design Engineering, Ulsan National 2 Institute of Science & Technology (UNIST), Ulsan, South Korea
E-mail: kmyung@unist.ac.kr

Abstract. Human-induced threats serve as potential hazards to cultural heritage assets, especially in developing areas where the local community, in general, is a deprived class. Sustainable tourism development is acknowledged as an economic activity to ensure careful management of assets along with local community empowerment and participation. As such, ICT-enabled development is applied in rural development projects to promote sustainable rural livelihood, but success is still limited due to a lack of community involvement and sharing in the economic gains of tourism. With this perspective in mind, the present study focuses on emerging marketing models (e-commerce) that can provide new business ventures for local communities by identifying critical online marketing elements driven by local residents.

1. Introduction

A large number of cultural heritage assets around the world are in danger from human-induced threats. Neglect, illegal trading of cultural property, theft, ethnic and religious conflicts, pollution, disappearances, fires, and ignorance all serve as potential hazards to heritage assets [1-4]. Some these human-induced hazards include the deterioration of the two 5th century Buddha statues in Bamiyan [5], the threat to the Taj Mahal posed by an oil refinery in India three decades ago [6], the neglect of the Buddhist Temple of Borobudur in Java that was rediscovered in 1814 under volcanic ash [7] and the Buddhist ruins in Takht-i-bahi in Pakistan [8].

Sustainable tourism development is acknowledged as an economic activity that ensures careful management of the use of inherited environmental and cultural heritage assets for preservation and tourism development [9, 10]. Because of geographic expansion and the labour-intensive nature of the tourism industry, there has been an increase in economic growth and a reduction in poverty in the developing world [11]. However, the developing world's share of international tourists is smaller in size and population when it comes to heritage tourism [12]. Heritage tourism is a subset of tourism that is utilized to increase regional development in rural areas [13, 14]. Heritage tourism entails places holding tangible and intangible features perceived by visitors [15]. Most heritage attractions are sponsored by elite groups or governmental bodies, and its representatives impose their own agendas and plans without involving local residents. It seems that local residents are entirely ignored in sustainable development processes, thus they become a deprived class and remain disempowered. Therefore, interpretation of heritage attractions should not be aimed mainly at visitors' motivations; it should delivered be more strongly to invite local residents [16].

Previous studies have emphasised local community empowerment through ICT-enabled development (e.g. [17], [26], [19], [12], and [20]). These studies have shown that empowerment is a voluntary, collaborative, and distributive process in which power and resources are used to enhance individual and collective capacities for sustainable economic and social development. Lyons et al. argued that community empowerment depends on commitment to empowering individuals through education and training [18]. Most local residents are poor farmers, landless workers, low-level traders, and rural artificers who are socially and economically deprived. As such, lack of participation in decision-making, of sharing in economic gains, and of awareness, funding, skills, and management to accommodate growth of heritage tourism turn into hurdles for sustainable development processes [21]. ICT-enabled development has been applied in rural development projects to promote sustainable rural livelihood by alleviating poverty through entrepreneurial activities and market access, rural empowerment and participation, environmental sustainability, and community network establishment [22]. To date, developing countries have faced considerable environmental issues related to the loss of natural habitats and biodiversity and threats to cultural treasures [23, 2]. In this regard, studies have focused on the use of ICT and Internet facilities to handle these issues through more environmentally sustainable models of economic development and to evaluate emerging eco-friendly technologies, equipment, and applications for sustainable development. However, most ICT projects face implementation challenges, and, ultimately, the use of ICTs results in projects that are not sustainable for rural and developing areas.

ICT-enabled development is categorized into two approaches. The first approach is the top-down professional development approach that pushes for local residents to participate via outside development [24]. Government agencies, NGOs, and other social workers provide financial support, leadership services, and technical assistance to empower local residents, fund micro-level entrepreneurship, and make new business ventures. In spite of such huge investments and resources and considerable trust of external bodies involved in sustainable rural development and livelihood enhancement, success has been limited [25]. The reasons for this may be deficiency, instability, or improper usage of financial resources; inexperience of external actors; and the need for local actors who can utilise local communities' skills for sustainable development. The second approach is community-driven development, whereby local residents drive their own development with full authorization and access to manage cultural and environmental assets without the feeling of being pushed into participation by external forces [26, 27]. This approach is more useful in terms of community participation because the top-down professional development approach, by not sharing economic gains, entirely ignores the readiness of local residents in the tourism management and development processes. However, the latter approach lacks the potential to create new business ventures and trade opportunities that generate net profits for the poor, covering economic, sociocultural, and sustainable environmental benefits rather than merely increasing revenue for businesses owned by third parties.

Most local residents are potential artificers and artisans who can reproduce antiques, such as cultural artefacts, paintings, costumes, and souvenirs that are significant, authentic cultural heritage assets [28-30]. Until now, the use of ICT-enabled development driven by local residents in antique reproduction has not been studied. With this perspective in mind, the present study focuses on emerging marketing models (e-commerce) that can provide new business ventures by identifying critical online marketing elements. For this purpose, this study first confirms an e-heritage e-commerce service for handicraft marketing using the Technology Acceptance Model (TAM), which was put forward by Davis [31]. The study concludes by suggesting a design for an e-heritage e-commerce ecosystem for handicraft marketing owned by local residents.

2. Research model

The objective of using TAM is to predict the probability that a new e-heritage e-commerce service for handicraft marketing will be adopted by participants who are potential customers (See figure 1). To determine the intention to accept the use of the service, two constructs, perceived usefulness and perceived ease of use, were used to predict whether the new service would be acceptable to potential customers. According to Davis et al. [32], intention is the degree to which a potential customer has expressed a disposition to execute or not to execute a particular future behaviour. This behaviour is affected by two main factors: usefulness and ease of use. For usefulness, customers may attempt to get many benefits of the e-heritage e-commerce service as per their interest in handicraft products. For ease of use, which concerns gaining the defined benefits of e-heritage, customers are able to easily do tasks such as collecting information and descriptions about cultural objects and their historical significance, ordering cultural artefacts of their choices, and educating themselves about cultural heritage assets.

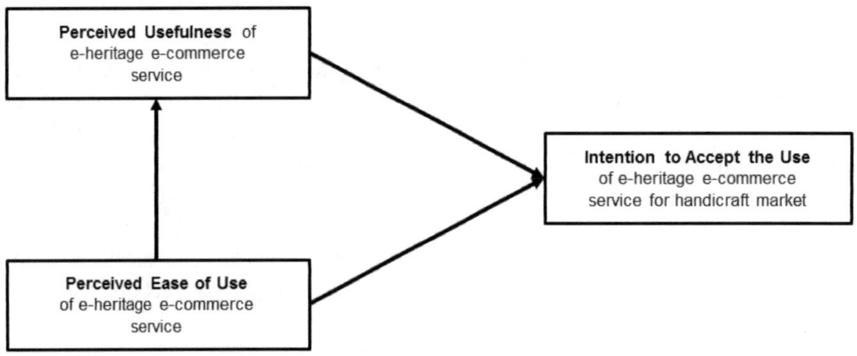

Figure 1. TAM for e-heritage e-commerce service.

According to this study, potential customers are those who find collecting/buying cultural objects over internet as useful and easy and who tend to accept the use of e-heritage e-commerce services for handicraft marketing. This suggests the applicability of using the TAM to predict the intention of potential customers to accept the e-heritage e-commerce service. This study seeks to answer the following question: How do perceived usefulness and perceived ease of use of e-heritage e-commerce services influence the intention of potential customers to accept their use?

3. Method

3.1. Questionnaire and constructs
The methodological guideline in this study was a quantitative approach specified to the TAM. The data used in this research was obtained through a standard questionnaire based on studies conducted by Davis [31] and Davis et al. [32]. The questionnaire was divided into three scales: perceived usefulness, perceived ease of use, and intention to accept the use. The majority of questions were adopted from previous research instruments (e.g. [33], [34], and [35]) with proven reliability and validity.

After modifying some questions, the reliability and validity of each construct of the research model (8 items for perceived usefulness, 4 items for intention to accept, and 8 items for perceived ease of use) was estimated by calculating Cronbach's alpha (α). None of the items were excluded, as the value of α fluctuated from 0.713 to 0.768 in all constructs.

The questionnaire's 20 items were measured on a 5-point Likert scale (from 1 = *strongly disagree* to 5 = *strongly agree*).

3.2. Sample
A total of 67 internet users were contacted via email. From that group, 43 users were considered potential customers (27 male and 16 female; age range 24–36), as they showed interest in cultural heritage and cultural artefacts. Because e-heritage e-commerce services for handicraft marketing are a new phenomenon, it is relevant to investigate users' intention to accept the use of the services.

3.3. Procedure and analysis

IOP Conf. Series: Journal of Physics: Conf. Series **989** (2018) 012009 doi:10.1088/1742-6596/989/1/012009

Based on current e-commerce principles, a dummy e-commerce store for handicraft products was developed. The products were taken from two rural areas of Pakistan: the Gandhara civilization in Taxila and the Indus Valley civilization in Harrapa and Mohenjo Darro. Both areas have historical significance and histories dating back to about 3300 BCE. As shown in figure 2, the store contains a total of 20 different products in seven categories.

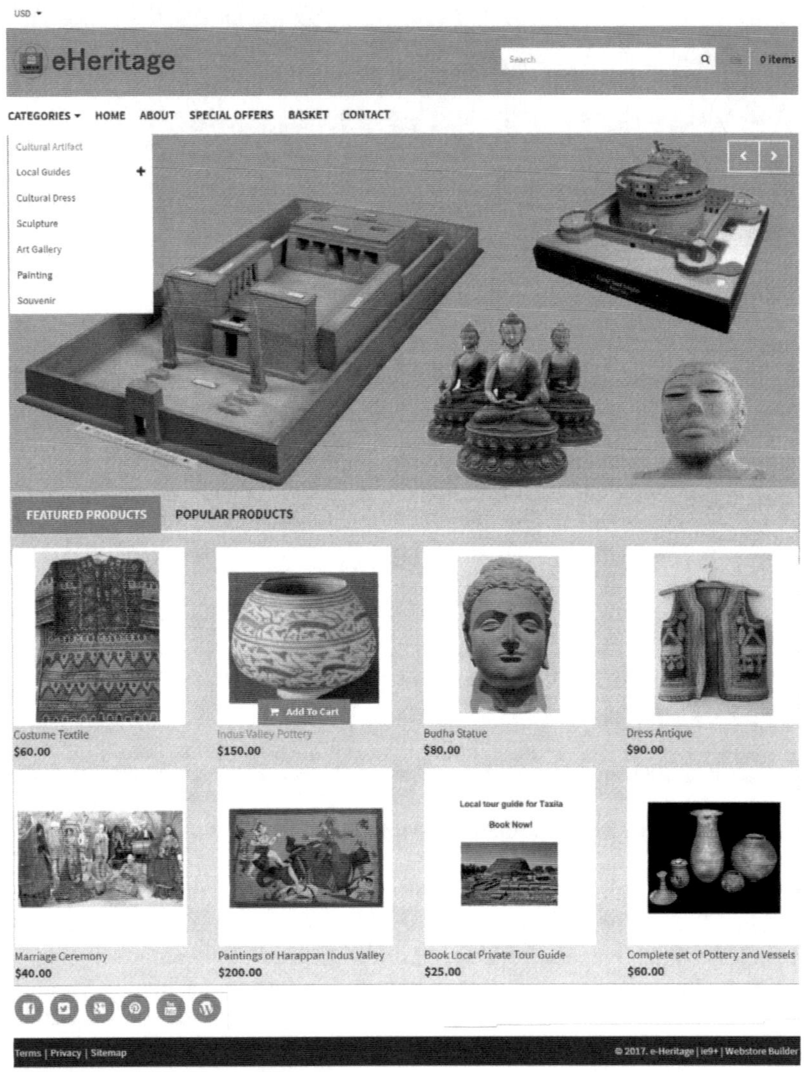

Figure 2. Main page of e-commerce store.

As shown in figure 3, users had to review each product and its description for its historical significance. Subsequently, the users had to respond to the questionnaire.

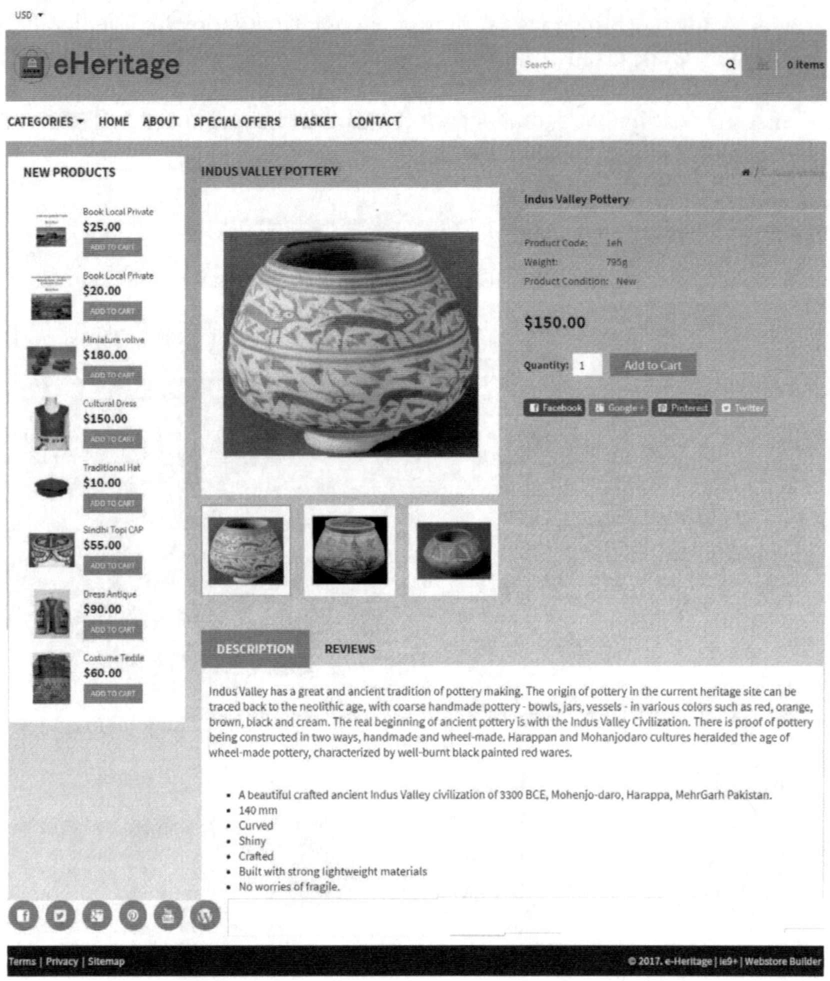

Figure 3. Product description and reviews page.

A multiple linear regression analysis was used to test the model for two independent variables – perceived usefulness (PU) and perceived ease of use (PEU) – and one dependent variable: intention to accept use (IEES). The objective was to determine the degree to which participants' readiness for online marketing benefits affected their acceptance of using an e-heritage e-commerce service for handicraft marketing.

4. Results and discussion

A multiple linear regression was performed to predict IEES for handicraft marketing based on PU and PEU. These variables' statistically significantly predicted the intention to use an e-heritage e-commerce service for handicraft marketing, $F\,(2, 40) = 41.18$, $p<0.0005$, $R^2 = 0.673$ (See table 1). The model interpreted 67% variance in acceptance of an e-heritage e-commerce service for handicraft marketing.

Table 1. Multiple linear regression analysis results

R Square	*F*	*Sig.*
0.673	41.18	0.000[a]
a. Predictors: (Constant), PU, PEU		

b. Dependent Variable: EES

In spite of the whole model's significance, each variable was tested for significance. Depending on this test, the main question was supported, including PU and PEU (see table 2).

Table 2. Constructs and regression results

Constructs	*Coefficients*	*T value*	*Sig.*
PU	0.433	2.23	0.031
PEU	0.355	2.03	0.048
Note. Significance level: $p<0.05$			

Both PU and PEU were significant. PU explained 43% of the variance and PEU explained 35%. This meant that 43% of the variance was accounted for by perceived usefulness and 35% by perceived ease of use, indicating the intention to accept the use of an e-heritage e-commerce service for handicraft marketing.

$$Model\ Compute\ IEES = (.433*PU) + (.355*PEU) - .615$$

The results confirmed the applicability of an e-heritage e-commerce service for handicraft marketing and explained the intention to accept use of one. The findings might be accredited to the nature of technological innovation in applying emerging marketing models in question. Because online handicraft marketing associated with cultural heritage assets is still new, there are currently not many customers, which might potentially explanation the inconsistencies. It cannot be argued that the potential customers are not aware of the presence of handicraft marketing, as all of the participants in this study showed interest in cultural heritage and cultural artefacts crafted by artificers. The findings showed that perceived usefulness was more influential in determining the use of an e-commerce service for handicraft marketing, thus it is crucial for designers in the field of sustainable development to design systems that are perceived to be more useful than easy to use.

5. Conclusions

It is obvious that a large number of cultural heritage assets around the world are potentially in danger from human-induced threats, especially in developing areas, because of local community disempowerment and deprived economic conditions. As such, sustainable development processes, including the sustainability of heritage assets, are seriously affected. Sustainable tourism development has been acknowledged as an economic activity that ensures careful management of heritage assets for tourism and community empowerment through ICT-enabled development. Studies have shown that the approach entails focusing on visitors' attraction to heritage tourism and its promotion while ignoring local residents, especially in the economic gains of tourism. This has resulted in deprived conditions and disempowerment for local communities. It has been suggested that needs of local communities should be elicited to consider the basic social and economic activities and that notions of sustainability should consider whether ICT approaches support activities that benefit local residents.

The present study focused on community-driven development to create new business ventures that generate net profits and that may empower local residents through active participation in development and economic gains. With this perspective in mind, this study presented notions of existing emerging marketing models that can be used to highlight local communities' skills in antique reproduction for new business ventures. It is widely known that local residents are potential artificers and artisans, so it suits them to bring forward their valuable skills through new, emerging markets. Thus, we developed an e-heritage e-commerce service for handicraft marketing by designing a website that contained cultural artefacts crafted by local artificers and artisans. Because the phenomenon is new, we used the TAM to confirm the intention of potential customers to accept its use. The findings showed the applicability of an e-heritage e-commerce service for handicraft marketing by determining the intention to accept its use.

The findings suggested the design of a self-organizing e-heritage e-commerce ecosystem that demonstrates community-driven development. The concept of an ecosystem in this context can provide a basis to underline local communities' skills and to reveal different roles, identifying internal actors in the community who can help create social and economic changes in developing areas that need sustainable development. In addition, such an ecosystem can provide a platform that gives rise to local residents who produce, supply, or distribute products that can be sold via e-heritage e-commerce services. In other words, local communities can easily participate in this type of service by constantly adapting the ecosystem. It can also be argued that an e-heritage e-commerce ecosystem for handicraft marketing could open participation of local communities by diversifying work opportunities. With these results, the study concluded by providing suggestions on how to utilize ICT-based community-driven development in developing areas with rich cultural assets by identifying new business ventures for deprived local residents.

Future research should continue to identify key actors that initiate, lead, and form the design of a self-organizing e-heritage e-commerce ecosystem. These actors may include internal actors in communities who identify and encourage craftsmen and local artificers, e-tailers and e-supply chain partners. These services need proper resources, support, and stakeholder engagement, thus institutional partners and sponsors should also be identified. Research should also continue under the assumption that such services can be sustainable and can foster sustainable heritage tourism development.

6. References

[1] UNECAP 2008 Cultural Tourism Sites Management – A Training Manual for Trainers in the Greater Mekong Sub region, United Nations, Economic and Social Commission for Asia and the Pacific http://lib.dtc.ac.th/ebook/Tourism/pub-2515-fulltext.pdf Retrieved 05/09/2017

[2] Nicolics Sandra, Laurent Richard, Helmut Jung and Reinhard Perfler 2014 Environmental Planning for Cultural Heritage Management—An Integrated Planning Approach for Heritage Site Communities *Regions Magazine* **293** 1: 11–14

[3] Gaast W Van Der 2017 A study on rural low-carbon tourism development in Jiangxi *IOP conference series* **1**–5

[4] Melat Aryasa, Alexander, Azis Nur Bambang, and Fuad Muhammad 2017 The study of environmental carrying capacity for sustainable tourism in Telaga Warna Telaga Pengilon Nature Park, Dieng Plateu, Central Java *In IOP Conference Series: Earth and Environmental Science* **70**: 1 012003

[5] Ashworth G J and van der Aa B J M 2002 Bamyan: Whose Heritage Was It and What Should We Do About It? *Current Issues in Tourism* **5** 447–57.

[6] Sharma J S and Sharma D N 1982 Atmospheric contamination of archaeological monuments in the Agra region (INDIA) *Studies in Environmental Science* **20** 31–40

[7] Gifford J 2011 Buddhist Practice and Visual Culture: The Visual Rhetoric of Borobudur Taylor & Francis

[8] Jokiletho J 2008 *The World Heriatge List: What is OUV? Defining the Outstanding Universal Value of Cultural World Heritage Properties*

[9] Fyall A and Garrod B 1998 Heritage tourism: at what price? *Managing Leisure* **3** 213–28.

[10] Zhao, Lei, Hongbing Zhu, and Shanfeng Hu 2017 Human-environment sustainable development of rural areas in China *In IOP Conf Series: Earth and Environmental Science* **64** : 1 012054

[11] UNWTO 2013 *Sustainable Tourism for Development Guidebook First edition*

[12] Timothy D J and Nyaupane G P 2009 Cultural Heritage and Tourism in the Developing World: A regional perspective *Routledge*

[13] Clark V 2007 Networks, technology and regional development: small tourism enterprises in Western Southland New Zealand Doctoral Diss. Auckland University of Technology

[14] Günlü E, Pırnar I and Yağcı K 2009 Preserving cultural heritage and possible impacts on regional development: Case of Izmir *International Journal of Emerging* … 1 13

[15] Poria Y and Ashworth G 2009 Heritage Tourism-Current Resource for Conflict *Annals of Tourism Research* **36** 522–5
[16] Marschall S 2012 "Personal Memory Tourism" and a Wider Exploration of the Tourism-Memory Nexus *Journal of Tourism and Cultural Change* **10** 321–35
[17] Sofield Trevor HB 2003 Empowerment for sustainable tourism development Emerald Group Publishing **7**
[18] Lyons M, Smuts C and Stephens A 2001 Participation, Empowerment and Sustainability: (How) Do the Links Work? *Urban Studies* **38** 1233–51
[19] Timothy D J and Boyd S W 2006 Heritage Tourism in the 21st Century: Valued Traditions and New Perspectives *Journal of Heritage Tourism* **1** 1–16
[20] Cole S 2006 Information and Empowerment: The Keys to Achieving Sustainable Tourism *Journal of Sustainable Tourism* **14** 629–44
[21] Scheyvens R 2002 Tourism for Development: Empowering Communities **27**
[22] Pade C I, Mallinson B and Sewry D 2006 An Exploration of the Categories Associated with ICT Project Sustainability in Rural Areas of Developing Countries : A Case Study of the Dwesa Project Africa **100**–6
[23] Houghton J 2010 Icts and the environment in developing countries: Opportunities and developments. The development dimension ICTs for development improving policy coherence: *improving policy coherence* **149**
[24] Kretzmann J P and McKnight J L 1993 *Building Communities from the Inside Out: A Path Toward Finding and Mobilizing a Community's Assets*
[25] Savitha V and Rajashekar H 2014 Evaluation of Major Problems Faced by the Members of Self Help Groups: A Study of Mysore District IMPACT: *International Journal of Research in Applied* **2** 59–64
[26] Leong C M L, Pan S L, Ractham P and Kaewkitipong L 2015 ICT-Enabled Community Empowerment in Crisis Response: Social Media in Thailand Flooding 2011. Journal of the Association for Information Systems **16** 1–39
[27] Mansuri G and Rao V 2004 Community-based and -driven development: *A critical review World Bank Research Observer* **19** 1–39
[28] Littrell M A, Anderson L F and Brown P J 1993 What makes a craft souvenir authentic? *Annals of Tourism Research* **20** 197–215
[29] Esperanza J S 2008 Outsourcing otherness: crafting and marketing culture in the global handicrafts market *Research in Economic Anthropology* **28** 71–95
[30] Hargrove C M 2017 Cultural Heritage Tourism: Five Steps for Success and Sustainability. *Rowman & Littlefield* **408**
[31] Davis F D 1989 Perceived Usefulness, Perceived Ease of Use, and User Acceptance of Information Technology *MIS Quarterly* **13** 319
[32] Davis F D, Bagozzi R P and Warshaw P R 1989 User Acceptance of Computer Technology: A Comparison of Two Theoretical Models *Management Science* **35** 982–1003
[33] Lund A M 2001 Measuring usability with the USE questionnaire Usability interface **8** 3–6
[34] Qiu L and Li D 2008 Applying TAM in B2C E-Commerce Research: An Extended Model Tsinghua Science and Technology **13** 265–72
[35] ALraja M N and Aref M 2015 Customer acceptance of e-commerce: Integrating Perceived Risk with TAM International *Journal of Applied Business and Economic Research* **13** 913–21

Acknowledgments
This work is supported by the Ministry of Education and National Research Foundation of the Republic of Korea (NRF-2015S1A5A8010614). The work is also supported by the Promotion of Special Design-Technology Convergence Graduate School of Korea Institute of Design Promotion (N0001436) with a grant from Ministry of Trade, Industry & Energy of the Republic of Korea.

IOP Conf. Series: Journal of Physics: Conf. Series **989** (2018) 012010 doi:10.1088/1742-6596/989/1/012010

Demonstration on Areca Catechu Tree Reuse with Supporting of Information Technology

F L Chao[1], C K Wu[1], A K Chao[2]

[1] Department of Industrial Design, Chaoyang University of Technology, Taichung 41349, Taiwan R.O.C.
[2] Department of Environmental Engineering, City College of New York, NY 10031, USA
E-mail: jkwu@cyut.edu.tw

Abstract. Areca catechu can be commonly found in Taiwan and Asia. By the restriction of agriculture policy, often the tree is chopped down and left in the wild and became an extra burden on the local environment. In this study, reuse design cases and opportunities were collected as Blog, so that people can access more easily. To enhance the user's awareness and information access it included the facets of its biology, culture history and reuse cases. Furthermore, we proposed demonstration supported with information technology. A blog can collect facts and examples with capabilities of multiple tags. This ability makes information search more accessible. The proposed approach combines both physical samples and visual elements in Blog which can be view by mobile phone. From the survey, Blog performs better than a regular internet search. Most people feel interesting, and some people were able to have own idea. Demonstration designs gather both elements will help to form a positive communication to the society with sustainable thinking.

1. Introduction

Areca catechu [1], often known as betel nut tree, can be commonly found in Taiwan around the rural or agricultural regions. It forms part of the unique scenery and culture memory. Though the specific ways consumption of its fruit, the betel nut [2], has been part of the local social history for years, the recent founding of its cancer-causing characteristic has resulted in its reduced consumption. By the agriculture policy, often the tree is chopped down and left in the wild. Part of the trunk fell into the river, also impede the smooth flow of stormwater and became an extra burden on the local environment (figure 1). Its plantation is also known to have a negative impact on the soil and water preservation. By identifying more about it, sometimes we found a different perspective. Though this is indeed true due to the tree's shallow root, however, it is not as bad as how the media portraits it. Recent years the government had been promoting to plant coffee trees along with areca trees to preserve the soil conservation.

By utilizing the specific properties or trait of catechu, it can be turned a burden into resources. Many products were designed using different parts of Areca catechu [3]. Recently, a facile synthesis of gold nanoparticles (GNPs) using the aqueous extracts of the aerial roots of Rhaphidophora areca intertwined over Areca catechu was carried out under different conditions, namely room temperature, higher temperature, solar irradiation and pH variation [4]. Apart from reducing waste, we can also get to know more about its unique growth pattern, and its close relationship with our ancestors.

IOP Conf. Series: Journal of Physics: Conf. Series **989** (2018) 012010 doi:10.1088/1742-6596/989/1/012010

(a) (b) (c)

Figure 1. Areca catechu and burden to the local environment (a) only head and leave cut-off, (b) trunk cut-off near the root, (c) section view of the trunk, from inside to outside, the outside made of hard Xylem and inner being vascular structure.

The primary intentions of the paper (figure 2) are : (1) collect knowledge related to the Areca catechu, (2) build a E-book Blog of Areca catechu, so that people can access more freely, (3) collect reuse cases (indicate by static demo in figure 2) related to the Areca catechu to promote designer to generate new design, (4) proposed a demonstration approach with both physical samples and visual elements to extend the impact to youth groups.

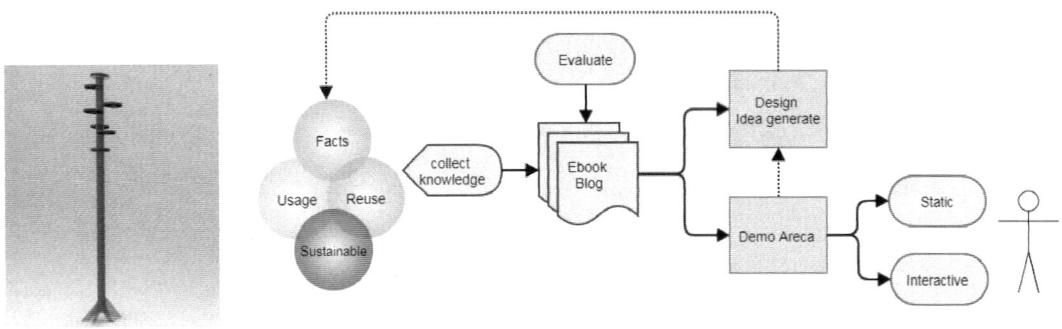

Figure 2. Main elements and intensions, physical display, acrylic tube contain samples and reuse design.

2. Collection of reuse designs

Our ancestor had a harmonic relationship with nature, plant and the soil. Areca catechu tree's unique composite material structure and its mechanism of transporting water are also an interesting phenomenon. While you find reuse opportunities, we can form a multi-faceted channel to pass information to our students and people. The E-book Blog and demonstration design discussed below, hoping that not only discovers reusability opportunities but also considers enhancing communication by providing the facets of its biology, culture history and sustainable facts. Design supported with information technology can extend the audience in a different location; cooperate to collect the idea and cases. The demonstration is undoubtedly helpful acquiring knowledge, experience, and feedback. That will support to form a positive communication loop with the society.

The areca catechu is broken down into four parts:

Leave: sheath, body and buds,

Fruit: ripe and unripe,

Trunk: from inside to outside, the outside made of hard Xylem and inner being vascular structure,

Root: radiant like; root body is round and there are enlarged nodes at the bending regions.

The design considerations included three facets. To better display different parts of the plant, we use a transparent acrylic pipe (figure 3). Different parts of the plant can show inside this pipe. The significant parts' characteristics are listed below (figure 4):
Leave sheath and body: Sheath reuse examples: fans, flip-flops and food plates. By extracting recycled sheath inner surface's fibers, it can form translucent stripes, which can be weaved into translucent films. The surface of a unripe nut is green. After cutting the nut it is edible. This is placed at the top of the acrylic pipe.

3. Physical samples design

Ripe nut gradually grows to a larger size. Its outermost layer becomes a hard, resistant fibers layer, and the middle core is a hard shell. By breaking the shell one can obtain the seed, which can employ as a dye. The central part of the acrylic tube displays products that made using this dye.
The trunk can be as tall as 20 meters. Its nodes are where the sheath is attached. The closer to the root of the stem the more it turns into xylem and phloem. The inside of the main trunk has the vascular bundle. Its xylem's texture is different to the one from ordinary trees. Each year it grows 3 or 4 extended rings. It is unique in that the vascular bundle growths in an oblique fashion. The slices made using a planer placed in the upper middle section so that one can observe that gradual change.
The cross-section of the trunk can be used to show its transition from hard to soft. We can join each acrylic pipe with a vascular bundle to demonstrate the water transferring behavior.
The root grows downward in an umbrella-like shape. A cross-section is made (added coloring) and placed in the acrylic pipe outer regions. The main trunk and the root reused as a substrate for flowers, or as unique patterned plates attaching to the ground.

(a)

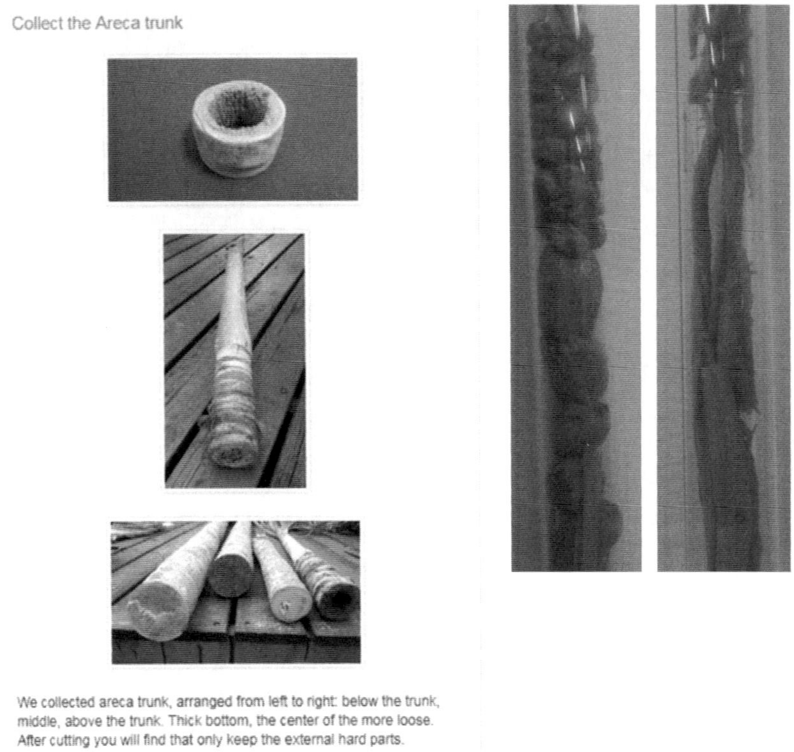

(b)

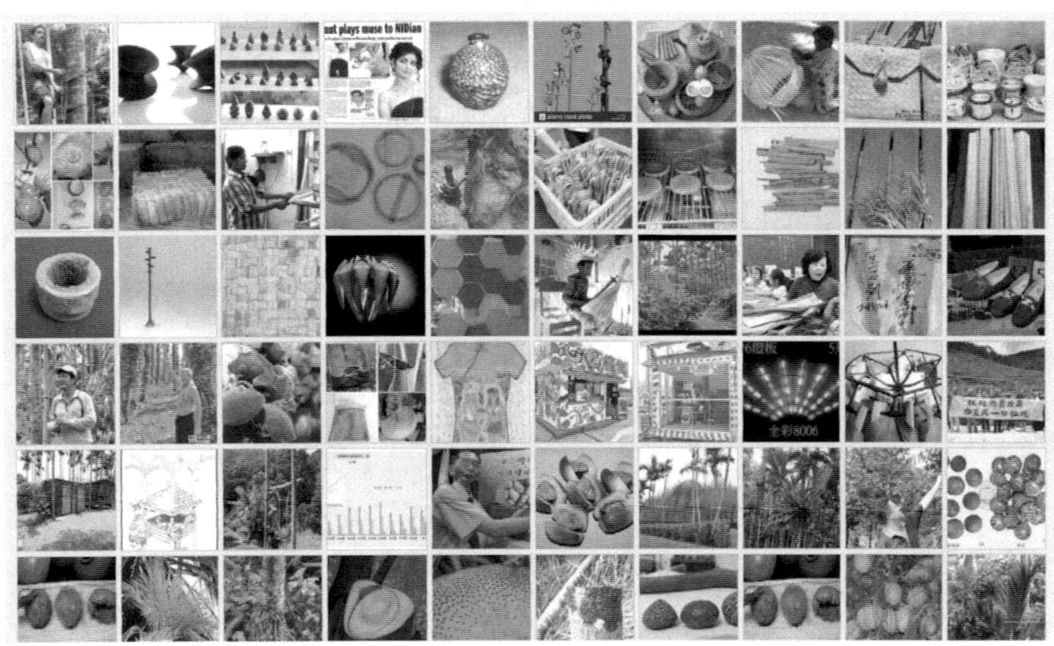

(c)

Figure 3. Samples of visual demonstration in Blog, (a) leading pages and corresponding tags (articles) on right side, (b) physical display, and (c) overall display mode.

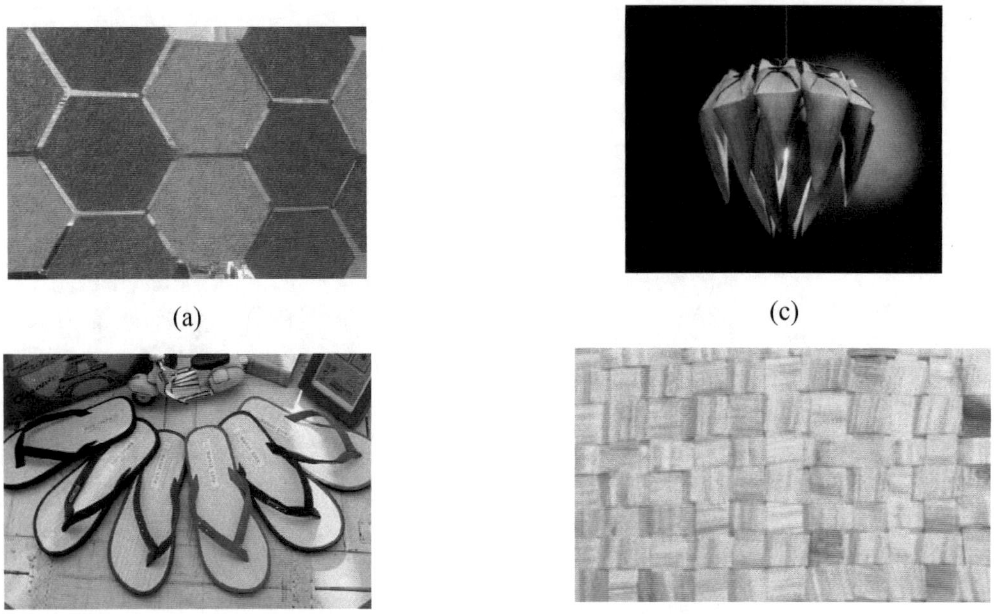

(a) (c) (b) (d)

Figure 4. Reuse cases of the different parts of the plant [5, 6], where (a) the carpet from unripe fruit (by T S Su), (b) the flip-flops (J L Lu) from portion of leave sheath, (c) light design (S Y Du) also by using leave sheath, (d) woven structure which collect from inner surface of sheath (S L Huang).

4. Demonstration supported with information technology

Scientific demonstration focus on information display which attracts people to further exploration. Knowledge-based society demands different attitudes than in the past regarding learning [7]. It is essential to develop, evaluate, and improve design using an iterative design-based research approach [8]. The idea is not only to give a representation of the facts, but the intention involves users in an interaction as part of the information access process.

Mobile devices extend the learning environment beyond classroom walls [9]. Mobile learning provided a new delivery mechanism to overcome traditional learning [10]. The concept of context-aware learning was identified as a learning environment that appropriate content can provide at the right time and in the right place [11]. By proper organize the physical and information design, it is possible to improve the possibilities of "provide at the right time and in the right place". Contextual information in the physical world and the adaptation accordingly to user action may provide customized learning content through mobile devices in response to situations. The trade-off between devices still needs to consider user's affordance.

We use Blog to describe the growth of the tree, the reuse procedure and reuse design details (interactive demo). The physical samples also included (static demo) to extend user's experience. When people see the physical display, they often request the information related to display examples.

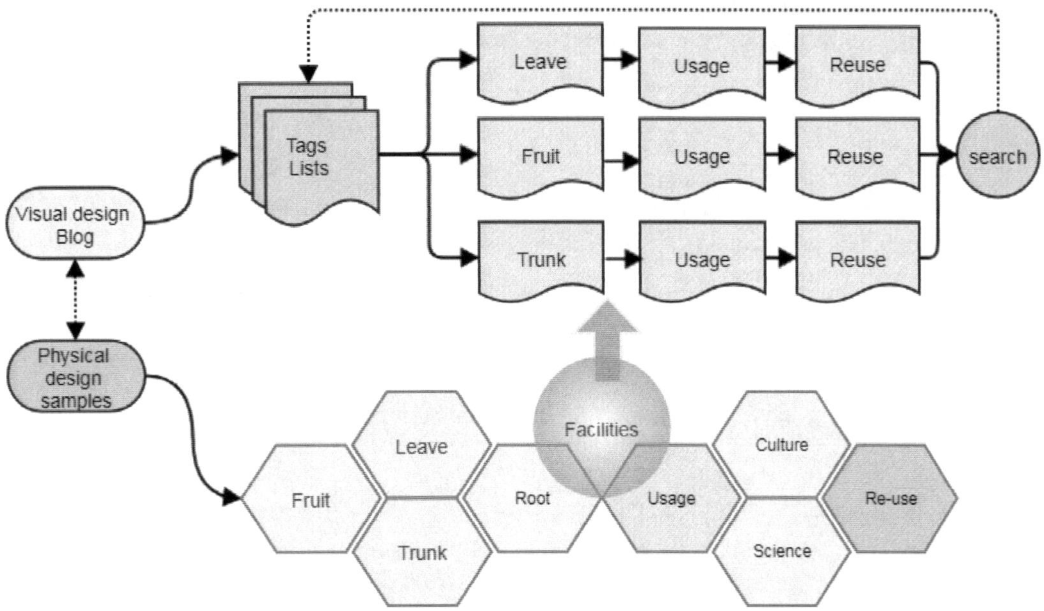

Figure 5. Communication design: upper part : e-book blog, lower part: physical facilities.

The Blog gives the visual demonstration, and acrylic tube display offers a physical presence. In the beginning, the audiences choose the part on the Blog that he or she wants to look. Within the Blog, everything saved in List with picture and text, and index points to different facet information such as cultural, plant structure, reuse and so on (refer lower part of figure 5). As Blog is demonstrating, the user likes to find corresponding on physical display, allowing the user to engage further. When the user is interested in a particular part of the material, he or she can search using the tags and trigger picture or text (refer upper part of figure 5).

E-book App was development firstly by choosing MIT App Inventor as the prototyping tool. The List structure utilized mostly for sequencing facts display, including picture/ text. The list index was organized as follow: 01-09 culture, 10-19 leave, 20-29 fruit, 30-39 trunk, 40-49 reuse of leave, 50-59 reuse of fruit, and 60-69 reuse of trunk. Depending on the collected data, the App will extend gradually.

Some difficulties were found with E-book App. There is no search ability, while the cases growth users cannot see what they want. MIT App Inventor has constraints on pages swap control and data retrieval with simple Tiny Database. Consequently, we shift our design from E-book App to E-book Blog.

The Google-supported blog application is Blogspot. It has become one of the top destinations on the internet to start a blog which is accessible with a Google account. Blogspot search engine that helps the user to find information focuses on the internal of the blog. The platform can collect facts and cases with capabilities of multiple tags. This ability makes information search much more comfortable.

5. Preliminary survey

Articles related to areca catechu tree are limited. We compared the user perceived depth on information regarding betel, between only use Google search to acquire knowledge of the subject, and experiencing our demonstrating design. Google search used as a reference baseline, where users conduct 15 minutes search and then answers a questionnaire. The facets of the questionnaire survey include:

- A: I think the betel nut tree re-use design is interesting.
- B: I acquired lots knowledge within 15 minutes using Blog.
- C: Sustainable design needs to understand a wide range of knowledge.

- D: Reuse design in different regions is similar.
- E: I can have ideas about reuse of betel nut trees.

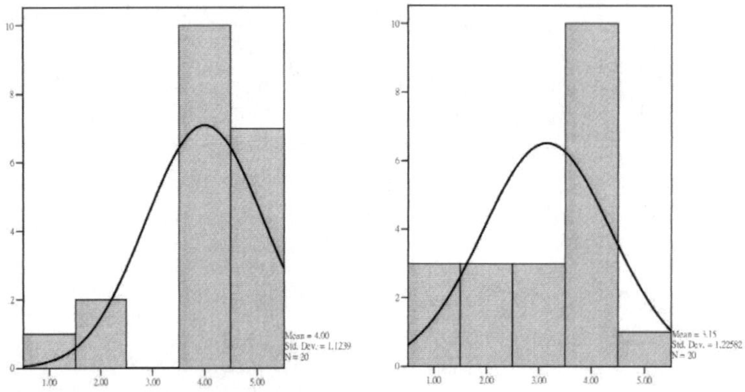

(a) (b)

Figure 6. Histogram results of question (a) A: re-use design is interesting and (b)E: have ideas about reuse, where vertical axis is the accumulated count.

The survey conducted on 20 users of Blog, it performs better than an internet search. Physical design supported with information technology extends positive communication with the society. Although internet search can provide a vast amount of information, they can often repeat. On the other hand, by using Blog, the audience can acquire facts through tags and systematic information structure within the Blogs. The averages (based on 5 point scale) of question A-E are 4.0, 3.7, 4.2, 3.3, and 3.15. By the message inspired, most people feel interesting. Some people have been able to provide an idea (figure 6). The demonstration approach with both physical samples and visual elements extended the impact to youth groups, and promote designer willing to generate a new idea.

6. Conclusion

Areca catechu is chopped down and left in the wild and became an extra burden on the environment. Re-look the Areca catechu can get an insight of reuse opportunities. People's awareness of its biology, culture history and scientific facts inspire personal intent of creating. The outcomes of the paper are: (1) collect knowledge related to the Areca catechu and build an E-book Blog of Areca catechu so that people can access more freely, (2) collect reuse cases related to the Areca catechu to promote designer to generate new design. From the preliminary survey, the Blog performs better than an internet search. The main demonstration elements collect diversify knowledge of the Areca catechu within the support of the extendable Blog. The collected reuse cases respire and promote designer to generate more new design. The demonstration approach with both physical samples and visual elements extended the impact of youth groups which gather local people participating in related events.

7. References

[1] Areca catechu, https://en.wikipedia.org/wiki/Areca_catechu
[2] Areca catechu – Betel Nut, http://entheology.com/plants/areca-catechu-betel-nut/
[3] https://udn.com/news/story/7266/2456169
[4] Jannathul M andLalitha P 2017 Cytotoxicity of spherical gold nanoparticles synthesised using aqueous extracts of aerial roots of Rhaphidophora aurea (Linden ex Andre) intertwined over Lawsonia inermis and Areca catechu on MCF-7 cell line, in *IET Nanobiotechnology,* vol. 11, no. 1, p 2-11
[5] https://www.damanwoo.com/node/87102
[6] https://bimlan.blogspot.tw/

[7] Marenzi I, Abel F, and Nejdl W, 2009 Social Sharing in LearnWeb2.0, *Int'l J. Continuing Eng. Education and Life-Long Learning,* vol. 19, nos. 4-6, p 276-290
[8] Wang F and Hannafin M J 2005 Design-Based Research and Technology-Enhanced Learning Environments, *Educational Research & Development,* vol. 53, no. 4, p 5-23
[9] Organero M M, Merino P J M, and Kloos C D, 2011 Adapting the speed of reproduction of audio content and using text reinforcement for maximizing the learning outcome though mobile phones, *IEEE Trans. Learning Technol.*, vol. 4, no. 3, p 233–238, Jul
[10] Chen C-C and Huang T-C 2012 Learning in a u-museum: developing a context-aware ubiquitous learning environment, *J. Comput. Educ.,* vol. 59, no. 3, p 873–883
[11] Hwang G-J, Tsai C-C, and Yang S J H, Criteria 2008 strategies and research issues of context-aware ubiquitous learning, *J. Educational Technol. Soc.*, vol. 11, no. 2, p 81–91

Toward Eco Product Development with Qualitative and CAE Design Process - Case Study of Flame Guiding Module

W L CHEN[1] and F L CHAO[1]

[1] Department of Industrial Design, Chaoyang University of Technology, Taichung, 436, Taiwan, R.O.C
E-mail: flin@cyut.edu.tw

Abstract. Sustainable products become increasingly important for company in addressing eco-performance to satisfy global environmental regulations. Case study of flame guiding module reviewed design process and concerns related to the torch design. For enhancing flame height, the torch was embedded with an airflow guidance structure. The design process and design methodologies were investigated as an eco-design case study. Combine qualitative and CAE simulation were proposed to fulfil its main and auxiliary functions including reduction of impact during use. The design guidelines help prevent mistake arrangements, CAE helps understand combustion phenomenon. The flow field simulation enables fine tune of geometric design. Functional test and measurement are carried out to confirm the product features. On Eco-performance, we choose 5 items for evaluation the status of previous and redesign module, namely function need, low impact material, few manufacturing steps, low energy consumption, and safety. The radar diagram indicates that eco-performance of redesign module is better. Life cycle assessment calculated the carbon footprint of the manufacturing and processing stage with Eco-it. By using recycled steel in the flame module, it reduces raw material stage carbon footprint significantly.

1. Introduction

Products eco-design aims to enhance usage, emotional value, and reduce the environmental burden throughout the entire product life cycle. Many eco-design methodologies based on life cycle thinking have been developed [1, 2]. Eco-design products are assessed based on functional, economic, and ecological criteria to achieve sustainability [3, 4].

1.1. Checklist

The checklist is one of eco-design methodologies [5] support for the analysis of a product's impact on the environment to avoid missing those features. It provides relevant questions that need to be asked about environmental bottlenecks. Usually, it starts with a needs analysis to fulfill its main and auxiliary functions. With this method, a designer can work out systematically redesign tasks on part, function and product levels. The steps include [5]: (1) define the existing product, (2) systematically answer the questions from the list, per stage of the product's life cycle, and (3) have improvement options as clearly as possible. The main strategies are (1) selection of low-impact materials, (2) reduction of material usage, and (3) modification of production techniques, and (4) reduction of impact during use.

1.2. CAE & Eco-design
The main target of flame module redesign is finding suitable parameters for enhancing natural convection. The heat transfer caused by fluids is expressed with the following equation:

$$Q_w = h\,(T_w - T_f) \qquad (1)$$

where T_w, T_f is the wall and fluid temperature. The heat transfer coefficient (h) depends on kind of fluid, its flow state, and the shape of the object. Computational fluid dynamics (CFD) techniques are foundation to predict airflow. Finite volume method (FVM) is a common approach used in CFD codes especially for large problems and source term dominated flows.

Sustainable future is an important driving force of innovation. The abilities of prediction and insight of problem make CAE simulation important in sustainable design. Designer used parametric optimization capabilities to design a car wheel hub that follows lines of force and avoids stress [6] with less material.

1.3. Qualitative guidelines
CAE does not necessarily solve the problem unless you find the right design. First of all, we need a basic understanding of the physical system, and then CAE detailed calculation steps support to get right parameters. Therefore, the designer also needs non-quantitative design guidelines.

For flow guidance design: the size and location of the opening shape need to be considered. Through the test and design changes, one can obtain the ideal airflow mode. Utilize the diversion port to guide the airflow in the ventilation path, reducing the occurrence of turbulence. Natural convection caused by negative pressure can bring into the air, to achieve an effective cooling cycle.

High surface temperature causes human risk. Careful considering the flow path can reduce the risk. The non-quantitative design guidelines can guide the initial concept of the proposal. The examples of the design guidelines are as follows [7, 8]:

- The mechanical elements must be demonstrated to a minimum useful lifetime of 20,000 hours.
- Thermal characterization should be carried out at the entire system level.
- Do not place the inlet of a blower directly downstream of the exhaust of another blower.
- Do not put two different blowers in series or in parallel unless the flow rates and head pressures in the system are balanced.
- Do not obstruct coolant flow, guide flow by means of splitters and turning vanes.
- Maximize the radii of turns in flow passages. Use gradual turns, enlargements, and contractions in flow passages.
- For free-convection, install cooling fins vertically. Do not place parts directly above high dissipating parts instead stagger them horizontally.

2. Case study of flame redesign
The case study reviews design process and concern on the torch. For enhancing flame height, the torch was embedded with an airflow guidance structure. Flame redesign process and design methodologies were investigated as an eco-design case.

2.1. Design procedure
Gas or Biofuel is chosen as fuel which is low oxygen consumption and capable of producing an ornamental flame with adjustable flame height. We built the digital model with CATIA solid modeling package. The digital model can be used to quickly build the tangible model by 3D printing. For simulation, digital model was discretized into small blocks (element), which can be set up the equation and boundary conditions. To solve the linear matrix equations, computer-aided engineering tool provide different solver depending on problem types, such as structural static stress or field distribution.

IOP Conf. Series: Journal of Physics: Conf. Series **989** (2018) 012011 doi:10.1088/1742-6596/989/1/012011

The prototype model can utilize for measuring actual physical parameters. The actual measured data and simulation results are compared to each other. We adjust the simulation parameters iteratively till meeting each other. When the parameters are confirmed, designer can use the digital model confidently to obtain design parameters. The major benefit of this development process is shorter cycle time through concept to the proof. Then manufacturing tradeoffs are also reviewed to confirm production feasible.

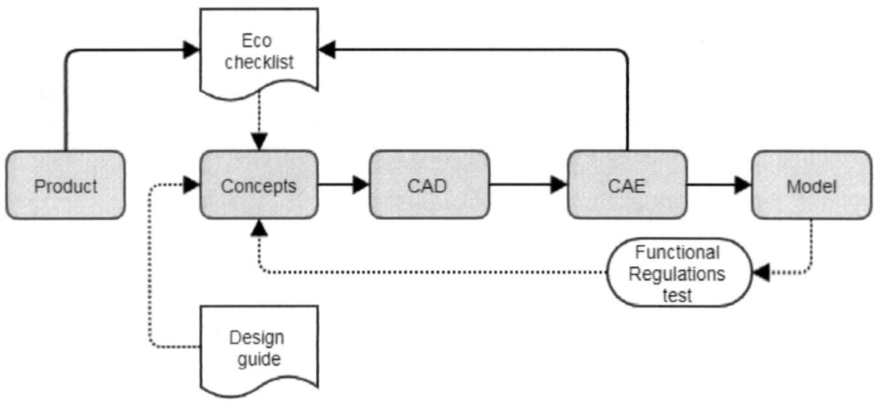

Figure 1. Design flow

As shown in figure 1, the design guide help prevent mistake arrangements. CAD/ CAE are used to quickly realize the idea. The flow field simulation enables fine tune of geometric design. Functional test proof and confirm the product features, through a series of measurements such as thermal energy, flame height, contact temperature (glass and metal). After that, the checklist was reviewed to enable the modified concepts.

2.2. Simulation results

In the case of the flame redesign, the complicated phenomena make CAE simulation much harder as described here. 3D surface and structure of the module are used to calculate the flow channel outlet speed, outlet angle and exhaust flow rate. Flow channel analysis was performed using non-premixed combustion mode. With the alcohol fuel combustion equation, through the amount of combustion to determine "Is the amount of oxygen supply is sufficient?"

$$C_2H_5OH + 3O_2 \rightarrow 2CO_2 + 3H_2O \qquad (2)$$

In non-premixed combustion, fuel and oxidizer enter the reaction zone in distinct streams [9]. This is in contrast to premixed systems, in which reactants are mixed at the molecular level before burning. Under certain assumptions, the thermochemistry can be reduced to a single parameter: the mixture fraction. The mixture fraction, is the mass fraction that originated from the fuel stream. In other words, it is the local mass fraction of burnt and unburnt fuel stream elements (C, H, etc.). Once mixed, the chemistry can be modeled as being in chemical equilibrium with the Equilibrium model, the Steady Laminar Flamelet model in which the microscopic element in the turbulent ensemble has the structure of an undisturbed laminar diffusion flame. The profiles of scalar properties provide an alternative thermochemical relationship to that of chemical equilibrium [9].

Analysis results of outlet angle and exhaust port velocity (figure 2) indicates the oxygen supply is sufficient. The XZ profile velocity chart (figure 3), the upwardly convection is on the outer edge of the cylinder. The streamline in overall flow line velocity diagram (Figure 3) is helically raised. Measured average temperature of the glass is 260° and that of outlet air is 505°. When the guide vane angle varies from 40 to 44 degrees, results of exhaust port velocity (table 1) indicated transition zone gradually downward shifted. Table 2 is suggested parameters based on simulation and measurement.

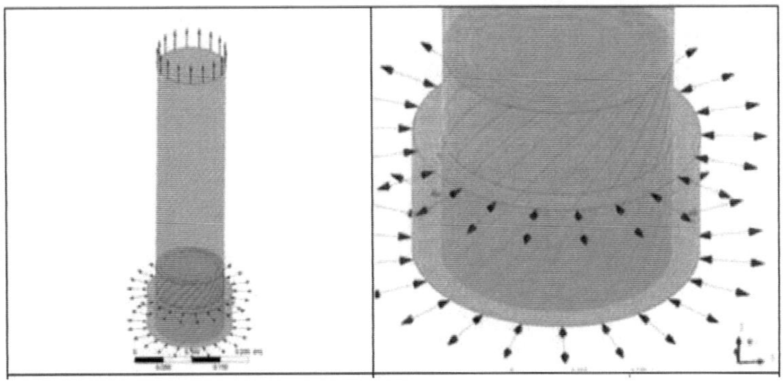

Figure 2. Non-premixed flames, modelling and inlet open boundary

Table 1. Analysis results of exhaust port velocity

Quantization value	entry value	Outlet value
Amount of airflow	3.1995(g/sec)	3.278(g/sec)
Oxygen content ratio	0.232	0.1948
Average speed	0.120(m/sec)	1.273(m/sec)

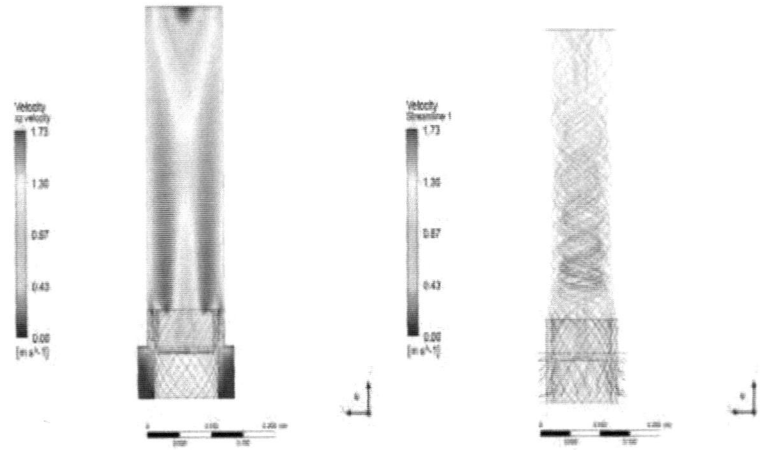

Figure 3. Outlet angle and exhaust port velocity analysis results

Table 2. Suggestions parameters

Variable	sensitivity	suggestions
Guide vane angle (30-60deg)	High	40
Glass tube height (30 to m)	Medium	36
length (3cm)	Low	>40
Number of leaves	Low (12-24)	20

IOP Conf. Series: Journal of Physics: Conf. Series **989** (2018) 012011 doi:10.1088/1742-6596/989/1/012011

2.3. Prototyping & measurement

The redesign of the flame height is much higher than the previous. Although the low diversion angle (below 40 degrees) caused outlet oxygen ratio slightly increased (complete flammability), the benefits are small (figure 4). Flame ornament ability decrease for the low angle, so the skew angle is recommended to be more than 40° . Prototyping models (figure 5) are used in the test to find better parameters. The main bodies of the ring and spiral material are steel, which was the high-temperature measured zone.

According to the regulations, we surveyed the most stringent CSA specification for the stove. There is spec. about the glass part of the open fire, but there is no high-temperature limit. Manufacture need warning mark on the glass. Thermal imager measurement showed the glass temperature of re-design is lower than related products. It revealed the effectiveness of the CAE design flow.

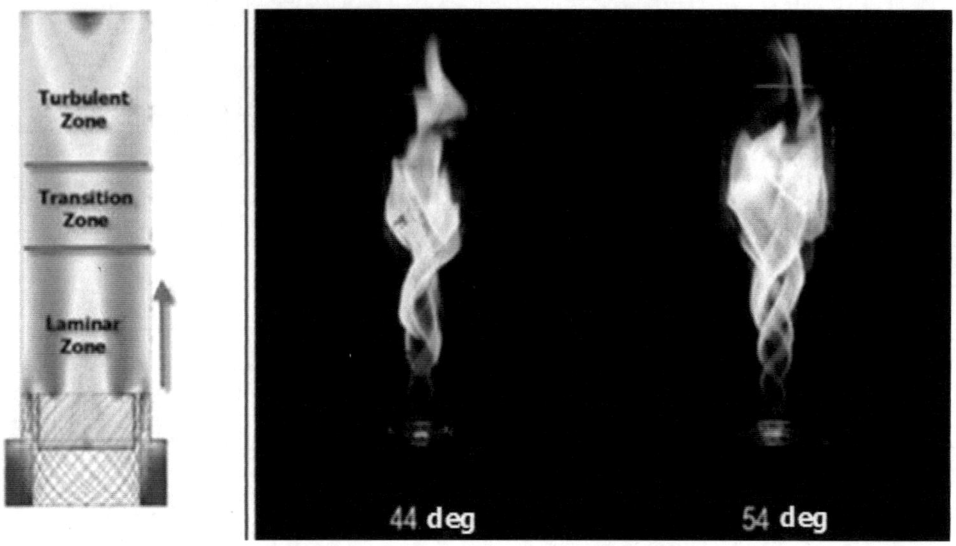

Figure 4. Measured flame for different skew angle of guide vane. Flame height is 35 - 44 cm, much higher than the previous design (15 ~ 22). Aluminium castings and iron plate metal, metal contact temperature is not higher than 60 degrees. Transparent parts: heat-resistant glass can be more than 600 degrees.

Figure 5. Prototyping model shown detail of the guide vane

IOP Conf. Series: Journal of Physics: Conf. Series **989** (2018) 012011 doi:10.1088/1742-6596/989/1/012011

3. Eco checklist and LCA

3.1. Checklist

According to [6], we select some keywords as follow: needs, fulfills its main and auxiliary functions, stage of the product's life cycle, options for improvement, low-impact materials, reduction of material usage, optimization, reduction of impact during use. By grouping related item, we choose 5 items for evaluation the status of previous and redesign module, namely function need, low impact material, few manufacturing steps, low energy consumption, and safety. The radar diagram shown in figure 6 indicates that eco-performance of redesign module is better than the previous product.

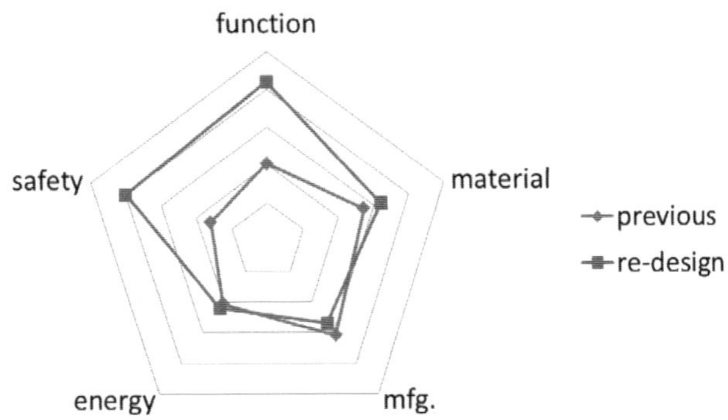

Figure 6. Eco-performance of redesign module

With a similar amount of material, redesign does not use the active device to push air and achieve substantial functional enhancement. Being appropriate combustion conditions, the redesign does not result in incomplete combustion of carbon monoxide. This is a significant reduction of the poisoning probability. No moving parts cause lower failure probability and product life increasing. All materials (figure 7) are recyclable also reduce the environmental impact.
Explore the reason of the significant improvement, for "fulfills its main and auxiliary functions", comes with the qualitative design guidelines and CAE on the combustion phenomenon. Design process helped fully explore the "options for improvement". Through the model test, modify and verify the simulation parameters, which enable "optimization" and achieve "reduction of impact during use".

3.2. LCA

Green design -the industry trend promoted low-carbon production, to enhance the competitiveness. We selected life cycle assessment (Eco-indicator 99) [10] using carbon footprint as an indicator, category from cradle to the door. The module of case study contains the raw materials (mining and manufacturing), transportation, processing stage till the module is completed [11], as shown in figure 7. Carbon footprints of the manufacturing stage are summarized with Eco-it in table 3 and 4. Flame module's total carbon footprint is 23.4 $KgCO_2$. If we the use recycled steel in future, it will reduce raw material stage carbon footprint significantly.

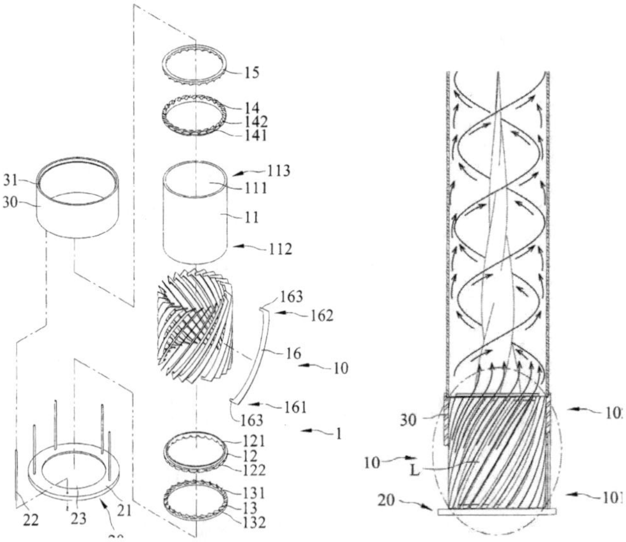

Figure 7. Parts of redesign module [12], where (16, 10) is flow guidance rib and channel, (L) indicate skew angle, (10) is module, and (20) is base frame.

Table 3. Raw material carbon footprint: 9.83 ($KgCO_2$)

Part No.	Name	Quantity	Weight (Kg)	Material	Footprint
1	Glass	1	0.921	Glass	0.51
2	Core	1	0.783	cast iron	1.16
3	Disc	1	0.402	White iron	1.9
4	Shell	1	2.082	cast iron	3.1
5	Wick	3	0.011	Stainless steel	0.16
6	Wick	1	0.053	White iron	0.25
7	Light body	1	0.108	White iron	0.51
8	Base	1	1.351	cast iron	2.0
9	Foot pad	4	0.02	rubber	0.24

Table 4. Processing carbon footprint: 13.6 ($KgCO_2$)

Part No.	Name	Quantity	Weight (Kg)	Material	Footprint
1	Glass tube	1	0.921	Annealing	0.168
2	Core	1	0.783	Casting casting	2.4
3	Disc	1	0.402	Stamping sheet	0.35
4	Shell	1	2.082	Casting casting	6.4
5	Wick	3	0.011	Coil, stamping	0.041
6	Wick	1	0.053	Inject molding	0.046
7	Light body	1	0.108	Inject molding	0.094

8	Base	1	1.351	Casting	4.10
9	Foot pad	4	0.02	Inject molding	0.014

Total carbon footprint in the processing stages 13.6 Kg. Transport phase Carbon footprint is parts weight (T) *Transport distance (km). The transport phase takes into account the carbon footprint of the parts delivered to the assembly plant using a truck. The distance was estimated by the shipped plan with Google Map. Due to the factory is cluster locate in Taichung area, total carbon footprint in the transport phase is 0.4.

4. Conclusions

The concept for sustainable products aims to enhance usage, emotional and economic value, and reduce the environmental burden throughout the entire product life cycle. It becomes increasingly important for company in addressing eco-performance to satisfy global environmental regulations. In this article, combine qualitative and CAE design process were proposed for fulfills its main and auxiliary functions and reduction of impact during use. The case study shows successful outcomes. Design process helps understand combustion phenomenon and fully explore the "options for improvement". Through the model test, modify and verify the rationality of the simulation parameters, engineers are able to select proper parameters and achieve "reduction of impact during use" .

References

[1] Ramani K, Ramanujan D, Bernstein W, Zhao F, Sutherland J, Handwerker C, Choi J, Kim H, 2010 Integrated sustainable life cycle design: a review *ASME J Mech Design* p 132

[2] Umeda Y, Takata S, Kimura F, Tomiyaama T, Sutherland J, Kara S, Herrmann C, Duflou J 2012 Toward integrated product and process life cycle planning - an environmental perspective. *CIRP Annals Manf Tech* **61(2)** p 681-702

[3] Arnette A., Brewer B, Choal T 2014 Design for sustainability (DFS): the intersection of supply chain and environment. *J Clean Prod* **83** p 374-390

[4] Spangenberg J, Fuad-Luke A, Blincoe K 2010 Design for Sustainability (DfS): the interface of sustainable production and consumption *J Clean Product* **18** p 1483-1491

[5] Kara S Honke I Kaebernick H 2005 An Integrated Framework for Implementing Sustainable Product Development, *Environmentally Conscious Design and Inverse Manufacturing 2005. Eco Design Fourth International Symposium* p 684-691

[6] 2017, www.moldex3d.com, Accessed: 10.10.2017. [Online] Available:http://www.moldex3d.com/en/newsletter/cae-simulation-technology-drives-opportunities-for-sustainable-automotive-manufacturing

[7] Morrison G N 1983 *Thermal guide for reliability engineer*, Hughes.

[8] Bahadur R and Bar-Cohen A. 2005 Thermal design and optimization of natural convection polymer pin fin heat sinks *IEEE Transactions on Components and Packaging Technologies*, ***28*(2)** p 238-246

[9] Liew S K, Bray K N, Moss J B 1981 A flamelet model of turbulent non-premixed combustion. *Combustion Science and Technology*, ***27*(1-2)**, p 69-73

[10] PRé Consultants, Goedkoop M, Oele, M, Schryver A, Vieira M, Hegger S 2010 *SimaPro Database Manual Methods library* **vol.2.4**, p 16-17

[11] Huang Shijie, Chen Jiahao 2015 Research on Carbon Footprint Evaluation of Mechanical Equipment Products, *Int. Conference on sustainability development* p B-1

[12] Chen W L 2011 *Taiwan patent* application no. 100100592.

IOP Conf. Series: Journal of Physics: Conf. Series **989** (2018) 012012 doi:10.1088/1742-6596/989/1/012012

A Study for Adjustable Riding Position of the Innovation Bicycle Design

Huang, Yu-Che [1] and Huang, Tai-Shen [2]

[1] Chaoyang University of Science and Technology, Bristol, Taiwan (R.O.C)
[2] Chaoyang University of Science and Technology, Bristol, Taiwan (R.O.C)
E-mail: che59965996@gmail.com[1], tshuang@mail.cyut.edu.tw[2]

Abstract. This research is about the analysis and design of the user in a wheelchair use posture can be changed. So user can through different sitting posture changes to provide different needs, so the folding design patents for the existing data collection and classification, through the analysis of advantages and disadvantages, further to find the innovative design approach. Finally through the multi linkage mechanism system, change the seat, handlebars and rear position, provide different posture context. In order to reduce the cost, this study is designed to integrate the innovation input mechanism, multi linkage system, through the construction of 10 joints and degrees of freedom of the bicycle frame. Then use 3D software to design of bicycle the size of the frame of innovation, through the three-dimensional CAD model to simulate the mechanism between movement and interference, and to evaluate the 3D The virtual prototype can meet the requirement. It will also enable CAE to analyze and determine the strength and the safety of the bicycle frame structure of ANSYS11.0 stress. Finally, we have integrated design through CAID to make a product appearance, the focus of research is mainly a study of CAD, CAE, CAID integration the production and the completion of a prototype. .

1. Introduction

Generally the wheelchair user in force are used in the upper part of the body, with the strength of the hand wheel to drive. But it is easy to cause people in the long-term use of shoulder joint and waist joint damage, because the force in a wheelchair on the handle only accounted for 25% of the body weight to 40%. This is why in recent years to a lot of Lun to changed structure similar to a bicycle, but the use of electric motor, at the same time to different wheelchair user needs, for example in the design of the bicycle, designed to provide different positions [1] horizontal bicycle system. It can reduce the burden of the body, can also provide a more comfortable environment this can be in the future. In order to study the design, the more rapid goods, and provide maintenance services in the future better, the research content will be with the bicycle frame as Design basis, change with multi link mechanism to guide the frame position. Provides two different postures, users can use a higher position in crowded urban areas to improve visibility, lift and can easily enter an indoor heat, and does not need to be outdoors after the move, need to replace vehicles to then enter the room, provide a more secure and comfortable environment. For a long distance ride, users can be replaced in the lower position, allowing the body to maintain a recline posture to reduce drag and improve the ride comfort. Of course, in the end, we try to use a simplified multi link mechanism to achieve the design goals, such as figure 1, as shown in figure 2.

IOP Conf. Series: Journal of Physics: Conf. Series **989** (2018) 012012 doi:10.1088/1742-6596/989/1/012012

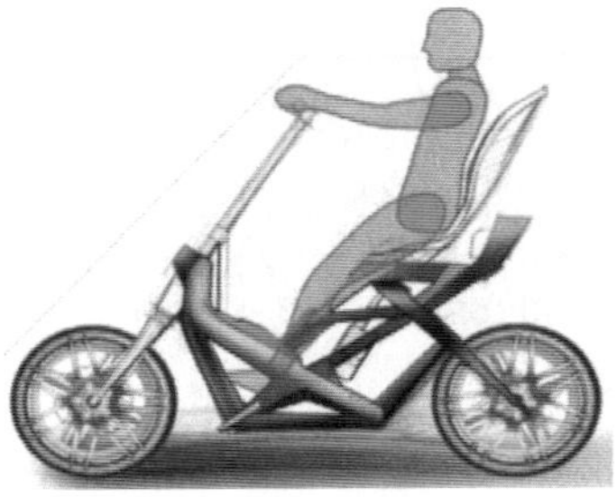

Figure 1.

Figure 2.

2. Literature review

This study will use for different situations were investigated, including research and design organizations fold related innovation patent, also can according to international regulations to do a complete discussion and integration can be a creative design content, then according to the different demand to the wheelbase design and development, innovation after the design is completed, the hope can provide the disabled with different needs, on the other hand can reduce the vehicle parking area, save space and increase road can be used. The most important thing is to provide users with comfortable environment and products that can be used in different locations, indoor and outdoor.

2.1 The variable-wheelbase bike that can narrow the parking space

This kind of type use P-Pair as the construction of folding bike chassis. The only machinery that qualifies this function is from Toyota Company's "i-unit future concept bike" which shown in 2004[2]. As shown in figure 3, it turns the machinery into skeleton picture. In figure 4 [3], the design concept is to view the bike frame as the combination of rigid body and linkage machinery for 10 bars, 13 joint and one freedom folding bike. Use P-Pair as the main connection is the bar 10 in figure 6. The design can drive a 4 bars linkages mechanism (bar 4, 8, 5, and 2) by the control of bar 10, and by this move can make the bike machinery to engage in the motion of folding or can get the same purpose by pushing the rear wheel slide part 9 to move to the front wheel. The other feature of this bike is there are 2 sets of R-Pair on the rear wheel slide part 9. It would do the rear wheel axle with a Row-way fixed function, so the rear wheel axle would become one freedom on the ground. The axle is basically qualified with great sustainability which is easy to install motor or battery component here.

Figure 3. i-unit future concept bike view.

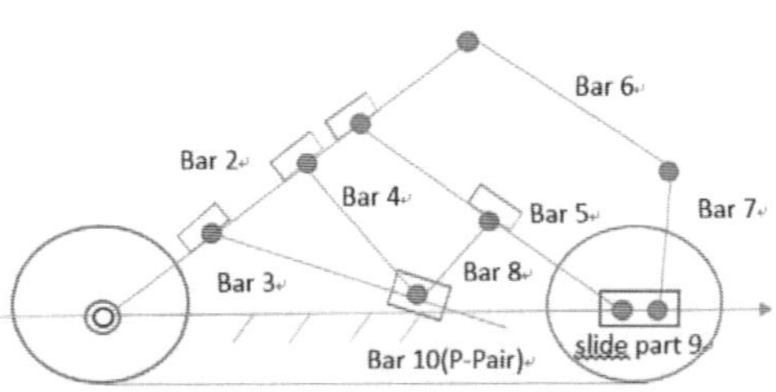

Figure 4. Linkage diagram.

2.2 The variable-wheelbase bike that can adjust riding position

In order to make it possible to commercialize the results of this study sooner, we will introduce the currently products available in the market when we embark on the establishment of actual restrictions and specifications for design, to make product viability fast. To enable adjustment of riding postures, our design allows adjustment of the distance between the shafts. As such, this study will set restrictions to product design with references made to the common bikes and recumbent bikes among the existing bicycle products.

Variable-wheel bike combines the merit of high efficiency from traditional bike, and the comfortable advantage of recumbent bicycle. However, in those two kinds of bicycle there consist of a big different in riding position, plus the harsh human size specifications in bicycle design and the rare document left in this field. In figure 5, it shows the bicycle frame construction of adjustable riding position. The patent use linkages set to lead saddle and frame construction moving, which can proceed the manual adjustment to the angle of saddle and the axle distance of wheels.

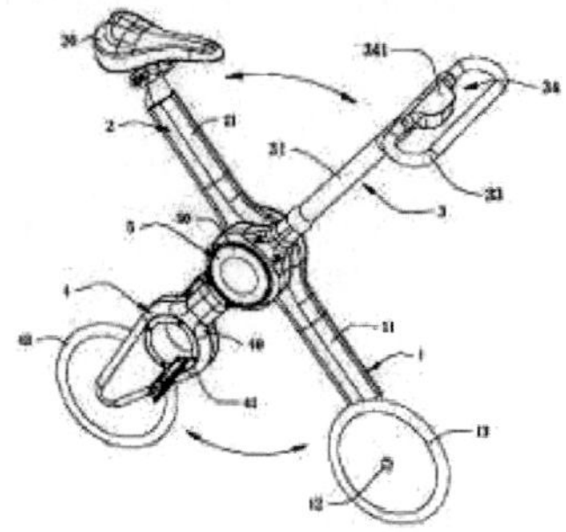

Figure 5. Patent exploded view.

3. Innovative design

This research take the bicycle frame as the design base and leading by multiple-linkages to alter the position of frame construction. User can adopt higher position when riding in the crowded urban area, which can raise the visibility and gives the user a safe ride. For long-distance riding, user can change for lower position in order to level down the wind resistance and step up for a more comfortable riding. We try to use simplified linkages to achieve the design goal. This study will also be through a variety of patents and research results, to do a concept and simulation analysis, and through the computer structure and strength of the analysis, and finally through the industrial design and sample production as a study of verification.

4. Conclusion

This study will be discussed and summarized through the relevant literature, and the relevant design and analysis, through the analysis of CAD and CAE, and then through the sample production to view the design differences and future changes in the direction, the contents of the design are as follows.

4.1 The specifications condition of variable- wheelbase bicycle

According to the user's findings and the collection of relevant laws and regulations, the following important design reference can be obtained:

(a) The design for front wheel steering more convenient, the innovation in the design of the front of the car with a wheel, and a motor, is behind the two wheels, a brake device, and in order to existing collocation commercially available products and convenient maintenance, so the tire diameter is 20 inches.

(b) This design allows the vehicle to maintain a complete comfort and stability during driving. The front wheel control angles must be the same at high or low riding positions, and maintain control capability.

(c) This design is in a high riding position; the rider's size must conform to the general frame geometry of the bike; at lower position it is necessary to apply the frame geometry to the reclining bike.

(d) This design is the use of a free linkage mechanism, using an electric actuator to drive the bicycle body structure into a higher or lower position when the handle seat height adjustment, the agency will also move, and seat
(E) This design on the change of riding position, the linkage system will automatically adjust the distance between the handlebar and the saddle, and the conversion process can be locked and fixed, and can move in the conversion process to provide fixed function [4, 5]
(f) This design has the same pedal position at high and low position, only the torso and thigh as thousands of angle change simplifies the design of repentance, body structure, and provide more secure to use.
(g) This design must be less than 400 millimeters across height and can touch the ground in any environment, which makes it easier for users to use. [6-8]
(H) This design changes the distance about 200 millimeters, under the long axis condition, the bicycle body lower part slope is biggest, is 25 degrees. (1) using electronic control design, including the design of lithium batteries, wheel motor drive. The electronic driving system controls the change of different sitting posture and adopts the general chain belt drive. (J) in order to qualify for the Taiwan electric bicycle standard, a bicycle without batteries should be light 40 kg, and the maximum speed should not exceed 25 km / h. [9-14].

4.2 Type synthetizing of variable-wheelbase bike
The linkage systems in this research must consider the higher and lower riding position before synthetizing. In order to maintain the same control ability before and after bike body transform, its angle between tube and ground need to sustain in steady value. So, the bike body and front wheel can be viewed as the fixed bar construction. The simplified design condition is more convenient in proceeding type synthetizing. We use 2 sets of 4 bars linkages to guide saddle and back wheel separately, and settle these two elements on the even number shaft of shaft group. Input and output shaft is connected with bike frame (fixed bar) and use the construction position synthetizing formula to calculate the length of 2 sets of linkages. With the same input angle variation, we can combine 2 input shaft with one rigid body, shown as figure 6 and figure 7, Watt-II one freedom.

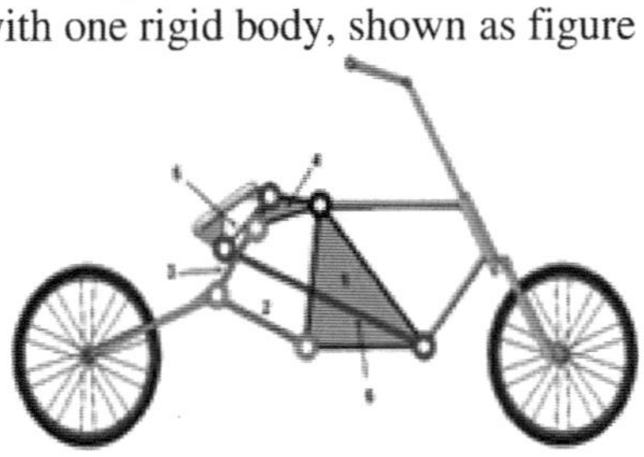

Figure 6.

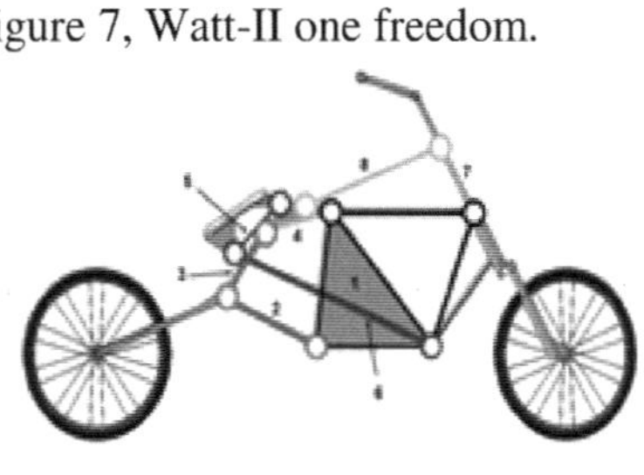

Figure 7. Adjust the connecting rod.

4.3 Size synthesizing of variable-wheelbase bike
After the research and analysis of patent information for the first step, we discover that the technology now can only reach to the part of partial adjustment of seat. We aim to improve the skill and design a newly light bike which is able to transforming completely into an innovative bike with the combination of "Backrest Type Bike" and "Lying Bike" , shown as figure 8 and figure 9 and figure 10. To get the right human size to qualify the comfortable appeal, we choose Giant Revive [15] as the higher position bike, whereas Bacchettabellandare_med is reference in design, shown as figure 11 and figure 12[16]. First, measure and draw the size picture of bike frame, and then overlap two pictures based on the design condition. Shown as figure 13 the reference bike frame, and the position shift and angle variation of saddle, pedal and handle bar. Then leading the reasonable human size design to make sure the research result is qualified to ergonomics. After conforming the angle variation, we can use the sport generator to calculate every shaft length which can achieve the goal of 3 4-bar Group using one input shaft collectively. Figure 14 is the completed picture of frame shaft length design. Figure 15 show diagrams for different riding positions.

IOP Conf. Series: Journal of Physics: Conf. Series **989** (2018) 012012 doi:10.1088/1742-6596/989/1/012012

Figure 8. Normal high sitting analysis.

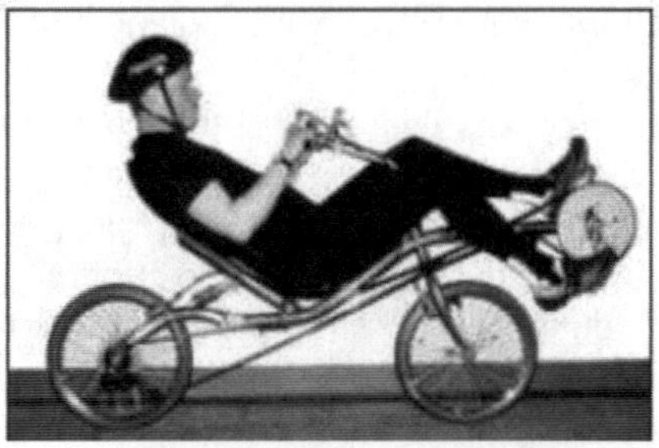

Figure 9. Normal low sitting image.

Figure 10. Different high and low sitting analysis.

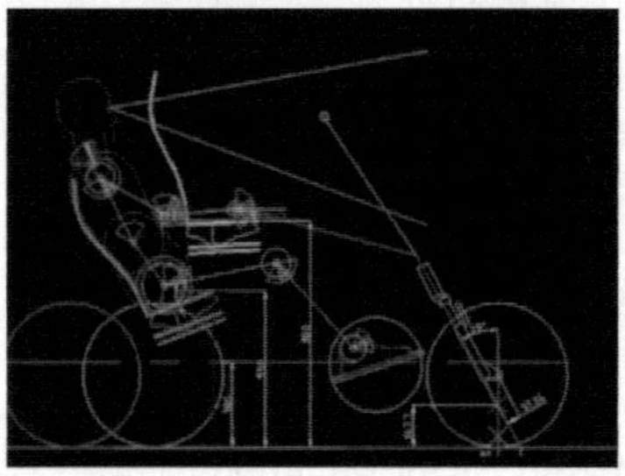

Figure 11. Different high and low sitting limit map.

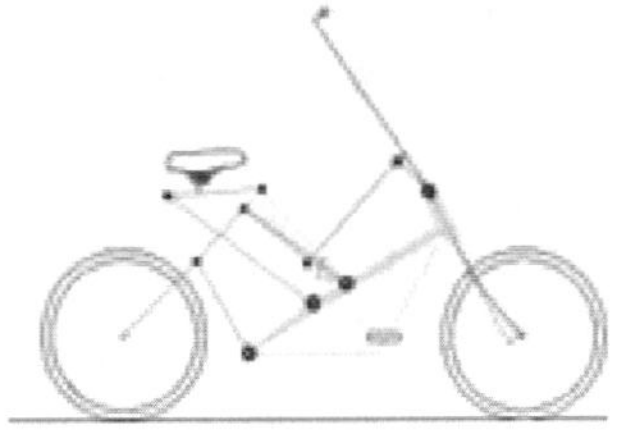

Figure 12. High sitting bar limit map.

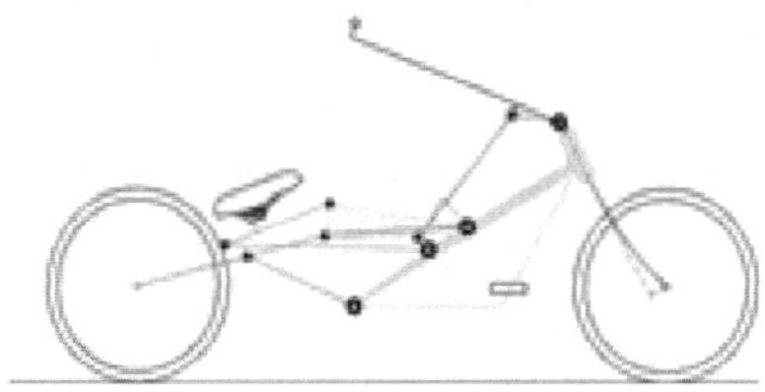

Figure 13. Low sitting angle limit chart.

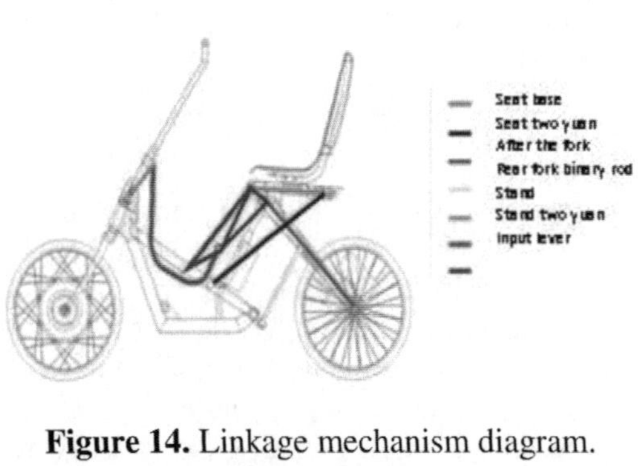

Figure 14. Linkage mechanism diagram.

Figure 15 Different sitting simulations.

IOP Conf. Series: Journal of Physics: Conf. Series **989** (2018) 012012 doi:10.1088/1742-6596/989/1/012012

4.4 3D model construction and application analysis of variable wheelbase bike

we use computer assisted design software SolidWorks to draw the 3D model, and design the detailed component size with the reasonable consideration of processing and cost. After completed element has already synthesized on computer, stimulating construction performance to make sure whether the component need interference or in the condition of couldn't fabricate. When the above condition has been fixed and affirmed, use value analysis software to proceed application analysis. Any defect in construction design which shown in figure 16 and figure 17 needs to change component material or alter the geometry construction in order to obtain completed security condition, and also we can make sure all the detail construction can perfume correctly. Lastly, we go into the product of rendering and animation shooting to 3D model, shown as Figure 18, so we can confirm the outer design before making into a prototype. Stimulating the control status of user and evaluate the future performance of the prototype. After a series of completed design and analysis, we go into processing production of all component output engineering picture, set circuit drive system, improve partial defect, complete the prototype machine, performance test, and confirm the original design idea of designers. Figure 19 shows the final product, including its exterior coating.

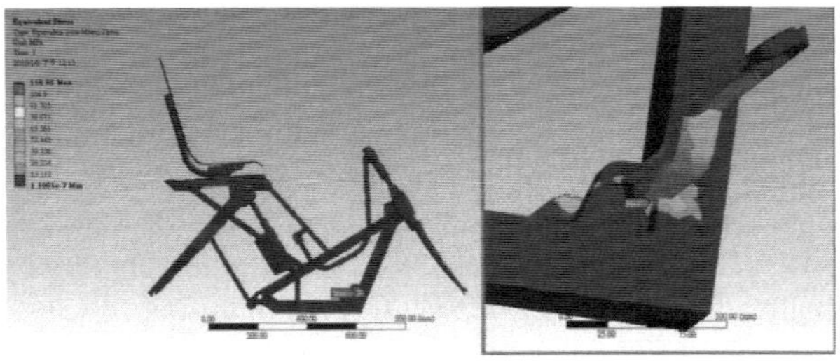

Figure 16 5 points force analysis.

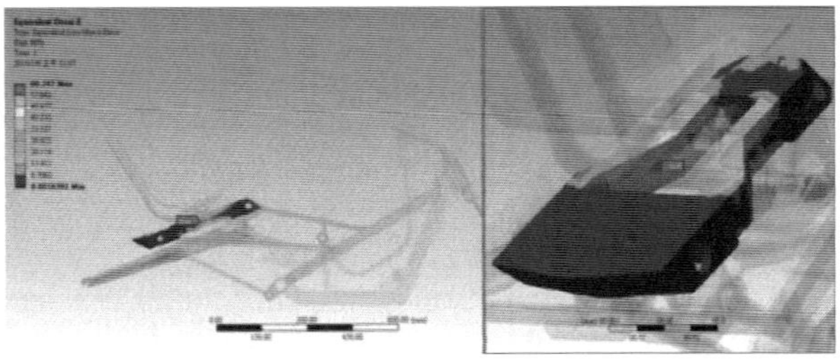

Figure 17 Seat force analysis.

Figure 18 Different sitting simulation.

Figure 19 The final product.

5. Conclusion

The research of the technology is the use of multi link to achieve a simplified shaft mechanism, further provide different design position, and provide a better control ability, including traveling in the outdoor and indoor, through the sitting posture change to meet the needs of the disabled. Through the relevant literature and other patent to make a comparison and analysis, through the analysis of the data for the design and development, can not only provide a method of sitting the requirement and design more and more can provide design and lower production cost. To provide a method of product design of the innovative design of the study, and the integration of different technologies to provide more innovative design and design method, the disabled can provide a better use of the environment, can also provide young designers a method of innovation integration design.

This is mainly through the design and research of the common bicycle parts to make a reference for the design, can provide faster and more convenient use, but also can reduce the weight of the product, through the multi link mechanism design, connection can help the different postures of institutions, a substantial increase in the variety, the design will be again through more intelligent, help the user's physiological response, further to provide better design, of course, these are must from the cognitive behaviors of users to study and observation, the future will continue to research and explore related.

6. Refererences

[1] C.Tai.Yu, Analysis and Design of Folding Car Mechanism"，Institute of Electromechanical Integration. National Taipei University of Science and Technology Master's thesis, Taiwan(2009)
[2] Mcocco, Giuseppe,"Adjustable- wheelbase Vehicle Chassis," E.P. Paten 1561674B1. (2007)
[3] Nmi Yasushi and Kayukawa Hiroshi, "Variable Wheelbase Type Vehicle, E.P. Paten 01721808A1.(2005)
[4] Li Yulong, Human Factors Design, Taipei: Taipei University of Science and Technology Course Lecture, (MAR.2008),pp. 1-2
[5] Frame size and geometric information sharing， http://www.frm.idv.tw/ redirect. php?fid=30&tid=3534&goto=nextoldset&sid=Uhr3TN
[6] Li.Bohao,The Effect of Bicycle Dimensions on Bicycle Performance")，Travel and leisure industry bimonthly，Vol. III, No. 7,(2003), pp. 18-34
[7] Kunlong, Special Topic Design Report of Industrial Design Department - Cruise Axis x Future. University Special Report, Chaoyang University of Science and Technology Master's thesis, Taiwan(2010)
[8] International Bicycle Design Competition Taiwan Bicycle Innovation Image System Platform Shift Bike，http://www. Ibdcaward. org/
[9] Nitute of labor safety and health，Human Measurement Database"，http://www.iosh. gov.tw/ Publish.aspx?cnid=26&P=828
[10] Recumbent bicycles Bacchettabellandare, http://www. oventrycycle.com/BentBikes.html

[11] Wei.Jiun.Huang, Jan.Nan.Chen, Exploration of Bicycle Design Trend from Grounded Theory－A Case Study of IBDC，Chaoyang University of Science and Technology Institute of Industrial Design Master thesis, Taiwan(2013)
[12] Tong.Chen,Jenq.Huey.Shyu,Research and Development of a Variable -wheelbase Vehicle.Taipei University of Science and Technology Institute of Electrical and Mechanical Integration, Taiwan(2010)
[13] Chun.Huang,Jenq.Huey.Shyu,Design and Analysis of Anti-theft Bicycles，Taipei University of Science and Technology Institute of Electrical and Mechanical Integration, Taiwan(2010)
[14] Expo Folding tricycle Di Blasi， http://www. nauticexpo.com/prod/di-blasi/folding- tricycle-22406-182240.html
[15] Stenberg, "Variable Geometry Cycle Frame," U.S. Paten 0212529.(2009)
[16] Ming-hui, and Stefan Palochev, "Bicycle frame structure with adjustable riding position"，ROC Patent Publication Number M417657. (2001)Examples taken from published papers

Sampling Based Influence Maximization on Linear Threshold Model

Su Jia[1] and Ling Chen[1]

1 Department of Computer Science, Yangzhou University, Yangzhou, China
E-mail: jiasu.forever@qq.com

Abstract. A sampling based influence maximization on linear threshold (LT) model method is presented. The method samples the routes in the possible worlds in the social networks, and uses Chernoff bound to estimate the number of samples so that the error can be constrained within a given bound. Then the active possibilities of the routes in the possible worlds are calculated, and are used to compute the influence spread of each node in the network. Our experimental results show that our method can effectively select appropriate seed nodes set that spreads larger influence than other similar methods.

1. Introduction

With the rapid development of Internet, people use social networks more frequently. In social networks, users can influence others' opinions, actions, emotion through their own influence propagation. The prevalence of social networks has brought up some new applications, for example: public opinion control, viral marketing etc. Hence, how to measure and forecast one user's or more users' influence on others in social networks is a challenging problem. The problem of influence maximization is defined as: given a network under a propagation model, we should find a set of starting nodes that can eventually influence the largest number of nodes in the network. The starting nodes set is called seed set, influence maximization will find the seed set with the largest propagation.

Recently, with the prosperity of many social networks, such as Facebook, Twitter, Sina and Renren, analysis of social networks has become a hot point of research. Especially the research on influence maximization in social networks has great significance in advertisement release, viral marketing[1], and message propagation.

Over the past decade, many new propagate models and influence maximization methods have been proposed. Those models and methods adopted theories and technologies in the areas of mathematics, computer science and data mining.

Kempe et al.[1] use a discrete optimization model to describe the problem of influence maximization. A given network is presented by a directed weighted graph, where nodes represent the users, and the weights on directed edges represent the influence between users. An integer k ($k \leq |V|$) is given as the size of the seed set, it is also called the budget. We need to find k users as seeds, and if they are activated and propagated under a diffusion model, the influence will get the largest spread.

In LT model [1], every node will choose a value from [0, 1] independently and uniformly as its active threshold. For any inactive node, when total weight of its active neighbors exceed its threshold, it will turn to be active. In both models, diffusion process will end when no more inactive nodes can be activated. At this point, the number of nodes that can be activated during the propagation process can be used to measure the influence of the seed set.

Domingos and Richardson[2], [3] first proposed and studied the influence maximization problem. They treat influence maximization as a discrete optimization problem, and propose a greedy algorithm. However, due to the huge time costed by the greedy algorithm, it can not be applied to social networks of large size. Leskovec et al. proposed CELF [2] along with its improved version algorithm CELF++ [3] and its improvement CELF++ which delay to calculate marginal gain of influence spread in order to reduce the processing time. Chen et al. [4] proposed a nodes' degree based method named DegreeDiscount which is an improvement of the Degree algorithm. Algorithm PMIA [5] constructs a tree structure for each node, and the influence spread and refresh is restricted within such tree region so as to reduce calculating time. However, it requires more memory space.
To reduce the computation time and the memory requirement, we present an efficient method for influence maximization on linear threshold model. The method first uses Chernoff bound to sample the routes in the possible world in the social network. After sampling, it calculates the active possibility of each route in the possible world. Finally, those possibilities will be used to calculate influence indicator of every node in social networks. Our experimental results show that the algorithm can effectively select appropriate seed nodes set to maximize the influence of propagation.

2. Concept and Definitions

2.1. Linear Threshold Model

A social network can be represented by a directed graph $G = \langle V, E, P \rangle$, where V is the vertex set, E is the directed edge set, $p = [p_{i,j}]$ is possibility matrix. Let $(v_i, v_j) \in E$ be a directed edge in G, element p_{ij} is the possibility for influence spreading from v_i to v_j At any time, each node in the network stays in either active or inactive state. An active node will stay in active state throughout the whole process. An inactive node can be activated by its active neighbors. The more active neighbors it has, the larger probability it can be activated. As one of the basic propagation models of social networks, linear threshold model [1] mainly focuses on the threshold behavior in the process of influence propagation, namely, the cumulative effect of influence during the propagation. In this model, when an attempt by an active node to activate its inactive neighbor fails, its influence will be accumulated during the active process. Such accumulated influence will have contribution to the possibility that other active nodes activate this non-activated node.Given a graph $G=(V, E)$, denote neighbors of node v in G as $N(v)$. Let S be the seed nodes set, $\theta_v \in [0,1]$ be the threshold of node v. The propagation process under linear threshold [1] is as follows. For a node v in G, denote the set of its active neighbors as $S(v)$. At any time, if $\sum_{u \in S(v)} p_{uv} \geq \theta$, then node v changes its state to active. After v being activated, it will attempt to activate its inactive neighbors. The influence of an inactive node will increase over time. The above procedure will be repeated until no more nodes can be activated.

2.2. Uncertain graph and possible world

We consider the original activation probability on an edge of the graph as the existence probability of the edge, then the graph can be treated as an uncertain graph.

Definition 1 (uncertain graph):

An uncertain graph can be represented by a triple $G = \langle V, E, P \rangle$, where $P: E \rightarrow (0,1)$ is the edge existence function, V and E represent the sets of vertexes and edges respectively.

The possibility $p \in (0,1)$ on each edge represents the existence possibility of this edge, if $p = 1$, then this edge definitely exists. If the probabilities of all edges in the graph are equal to 1, the graph is a traditional deterministic graph.

Definition 2 (Possible World):

For an uncertain graph $G = \langle V, E, P \rangle$, if a certain graph $G' = \langle V, E' \rangle$ satisfies $E' \in E$, then G' is one possible world that belongs to G.

Obviously, an uncertain graph $G = \langle V, E, P \rangle$ contains $2^{|E|}$ possible worlds. We denote the set of all possible worlds of G as $W(G)$.

Let $G' = \langle V, E' \rangle$ be a possible world of G, then the existence possibility of G' is:

$$\Pr(G') = \prod_{e \in E'} p(e) \prod_{e \in E - E'} [1 - p(e)] \qquad (1)$$

It can easily be seen that the sum of existence possibilities of all possible worlds of G is 1, that is:

$$\sum_{G' \in W(G)} \Pr(G') = 1 \qquad (2)$$

3. The Activation Possibility of Route

Kempe etc.[1] had proved that LT model is equivalent to the "Live-edge" model (LE) [1].

In LE model, every node chooses one in-edge as "live-edge" based on the probability of in-edge connecting with the node.

Once an edge is selected, it can be called a 'live' edge, while the other edges not selected are called 'dead'.

After such edge selection for all the nodes, we can obtain a possible world of G. Let $\sigma(S)$ the number of nodes that can be activated by the seed set S in the LT model. Then $\sigma(S)$ is equal to the mathematical expectation of the number of vertices that can be reached in all possible worlds.

Denote G' as a possible world of G, $\sigma_{G'}(S)$ is the number of reachable nodes starting from S in Graph G', then we have

$$\sigma(S) = \sum_{G' \in W(G)} \Pr(G') \sigma_{G'}(S) \qquad (3)$$

Then $\sigma_{G'}(S)$ can be calculated by the following formula:

$$\sigma_{G'}(S) = \sum_{v \in V} I(S, v, G') \qquad (4)$$

In (4)

$$I(S, v, G') = \begin{cases} 1 & exits route from\ S\ to\ v \\ 0 & otherwise \end{cases}$$

Combine (3) and (4) we can get

$$\begin{aligned}\sigma(S) &= \sum_{G' \in W(G)} \Pr(G')\sigma_{G'}(S) \\ &= \sum_{G' \in W(G)} \sum_{v \in V} I(S, v, G')\Pr(G') \\ &= \sum_{G' \in W(G)} \sum_{u \in S} \sum_{v \in V} I(S, v, G')\Pr(G') \\ &= \sum_{u \in S} \sum_{v \in V} \sum_{G' \in W(G)} I(S, v, G')\Pr(G') \\ &= \sum_{u \in S} \sum_{v \in V} \Pr(u, v) \qquad (5)\end{aligned}$$

Here

$$\Pr(u, v) = \sum_{v \in V} \Pr(u, v)$$

is the activate possibility of the existence of the route between u and v.

Now we define:

$$\Pr(u) = \sum_{v \in V} \Pr(u, v) \qquad (6)$$

then

$$\sigma(S) = \sum_{u \in S} \Pr(u) \qquad (7)$$

From (7) we can see that calculating $\sigma(S)$ needs to compute $\Pr(u)$ for every node u in S, namely, we should calculate the possibilities of all the routes starting from u.

4. ROUTE SAMPLING

The use of sections to divide the text of the paper is optional and left as a decision for the author. From (6) we can see that formula (7) needs to calculate $\Pr(u, v)$ for every vertex v, where $\Pr(u, v)$ is defined as the existence possibility of the route between u and v, that is:

$$\Pr(u, v) = \sum_{G' \in W(G)} \Pr(G') I(u, v, G') \qquad (8)$$

It can be seen from formula (8) that computing $\Pr(u, v)$ needs to calculate $I(S, v, G')$ in all possible worlds of G. That means to detect whether there is a path from node u to v in every possible world. It needs to enumerate all the possible worlds of G, which requires huge amount of computation time. To reduce the computational cost, we use a sampling method to randomly select some possible worlds from $W(G)$. Denote the set of selected possible worlds are $U(G)$. We use $U(G)$ as a replacement of $W(G)$, and use

$$\frac{1}{|U(G)|} \sum_{G' \in U(G)} I(u, v, G')$$

as an unbiased estimation of

$$\frac{1}{|W(G)|}\sum_{G' \in W(G)} I(u, v, G')$$

To restrict the error of the estimation with a given bound ε, we use Chernoff bound to estimate the size of sampling set $U(G)$.

Theorem 1 (Chernoff Bound):

Let $X_1, X_3, \ldots, X_r$ be independent and identically distributed random variables, they have the same exception $\mu = E[X_i]$, if $r \succ \frac{3}{\varepsilon^2 \mu}\ln\frac{2}{\delta}$, then

$$\Pr\left(\left|\frac{1}{r}\sum_i X_i - \mu\right| \geq \varepsilon\mu\right) \leq \delta$$

Let $I_1, I_2, \ldots, I_r$ be the values of $I(u, v, G')$ of the possible worlds $G_1, G_2, \ldots, G_r$ in $U(G)$ and $U(G) = r$, then

$$\mu = \frac{1}{|W(G)|}\sum_{G' \in W(G)} I(u, v, G')$$

is the exception for $\frac{1}{r}\sum_i i$. For a given probability δ, if we set

$$|U(G)| \geq \frac{3}{\varepsilon^2 \mu}\ln\frac{2}{\delta} \qquad (9)$$

Then the relative error of estimation will be less than the given error bound ε.

Since the exception $\mu = E[X_i]$ is required in (9), we set a threshold ρ such that the route (u, v) with $\Pr(u, v)$ less than ρ can be ignored. Then we can use ρ to replace μ in (9), and set

$$r = \frac{3}{\varepsilon^2 \rho}\ln\frac{2}{\delta} \qquad (10)$$

as the size of sampling set.

5. Framework of The Algorithm and Complexity Analysis

According to the above analysis, we propose a sampling based influence maximization algorithm named Sampling_IM. First, the algorithm calculates the sampling size r using (10), and constructs r possible worlds. Then it calculates the activate routes in each possible world. Based on the activate routes obtained, the algorithm can compute the route's activating possibility $\Pr(u)$ for each node u. Finally, we select p nodes with the largest $\Pr(u)$ values as the seeds set S. Given an error bound ε, possibility δ, route exception threshold ρ, the framework of our algorithm Sampling_IM is described as follows:

Algorithm1 Sampling_IM
Input: $G = \langle V, E, P \rangle$

p: size of seed set;
ε: error bound;
δ: possibility (confidence);
ρ: routes exception threshold;
Output: S: seed set;
1:use (10) to calculate sampling size r;
/*construct r possible worlds*/
2: for k=1 to r do
construct_PW(G,P,G_i);
end for k;
3: for all node pair (u,v) in G do
Pr(u,v)=0;
end for;
/*calculate the number of nodes that can be activated by each node u of the graph */
4: for i=1 to r do
path_counting(G_i);
end for i;
5: for every node $u \in V$ do
$\Pr(u) = \sum_{v \in V u \neq v} \Pr(u, v)$;
end for u;
6:get p nodes with larger values of $\Pr(u)$ to formulate seed set S;
7:**output** seed set S

In algorithm Sampling_IM, step 1 calculates the sampling size r using (10), given parameters ε, δ and ρ. Step 2 constructs r possible worlds of graph G. Steps 3 and 4 calculate the number of nodes that can be activated by each node u in the graph. Step 6 selects p nodes with the larger $\Pr(u)$ values than that of the others to constitute the seeds set S.

In the algorithm, step 1 uses (10) to calculate r, the complexity is $O(1)$. Step 2 constructs r possible worlds of graph G, the complexity is $O(r*m)$, where $m = |E|$ is the number of edges in G. The complexity of Step 3 is $O(n^2)$.

Step 4 calculates the number of nodes that can be activated by each node u in G, its complexity is $O(r * n^2)$. Step 5 accumulates the $\Pr(u)$ for all nodes in G, it costs $O(n*p)$ time. Step 6 selects p nodes to constitute the seeds set S, its time complexity is $O(n*p)$. Step 7 outputs seed set S in $O(p)$ time. In summary, the complexity of algorithm Sampling_IM is $O(n^2)$.

6. Experiment and Analysis

In order to verify the accuracy of the proposed algorithm, we conduct experiments on four datasets DBLP, Epinions, LiveJournal and NetHPET that are benchmarks in the literature of influence maximization[7]. Then we analyze the test results.

The algorithm is coded by Matlab and run on a PC with 2.4GHz Intel I5 6300HQ, 16G memory under Microsoft Windows 10.

6.1. Datasets and contrast algorithm

We test our algorithm on three true world datasets as shown in TableI.

TABLE I. EXPERIMENT DATASETS

Name	DBLP	Epinions	NetHPET
Node	650k	13k	15k
edge	2M	84k	62k

We compare our algorithm with PageRank [6] and DegreeDiscountIC [4]. PageRank is a commonly used algorithm for ranking web pages. In LT model, the transition probability on edge (u, v) is interpreted as influence of node v on u. After the nodes are ranked by PageRank, k nodes with the highest ranking values are selected as the seed set. DegreeDiscountIC is used for IC model. In the method, all edge propagation probability is set as 0.01. The basic idea of this algorithm is that it always activates the inactive nodes with the largest degree. However, after some neighbors of node v are selected as seeds, the degree of v should be decreased by a discount. Then the algorithm activates v according to its discounted degree. In LT model, this method can be used as a general heuristic method [4].

6.2. Propagation model and settings

Linear threshold model (LT) is a classic propagate model, we widely used WC method [1] to set up parameters on the edges.

The propagation possibility on edge (u, v) in graph G is set as:

$$b_{u,v} = \frac{A(u, v)}{N(v)}$$

Here, $A(u, v)$ is the number of directed edges from node u to node v, and $N(v)$ is the indegree of node v:

$$N(v) = \sum_{u \in N^{in}(v)} A(u, v)$$

The activate threshold is generated randomly and uniformly between 0 and 1.

6.3. Experiment result and analysis

Tables should have only horizontal rules and no vertical ones. Generally, only three rules should be used: one at the top of the table, one at the bottom, and one to separate the entries from the column headings. Table rules should be 0.5 points wide.

6.3.1. Result on DBLP. In the experiments on DBLP, we set $\varepsilon = \frac{1}{100000}$, $\rho = 6$, $\delta = 0.99$ By (10) we get $r = 3516000$ which means that we should sample 3516000 possible worlds in DBLP in our experiment. We compare the influence spreads of different algorithms, and the result is shown in Figure 1.

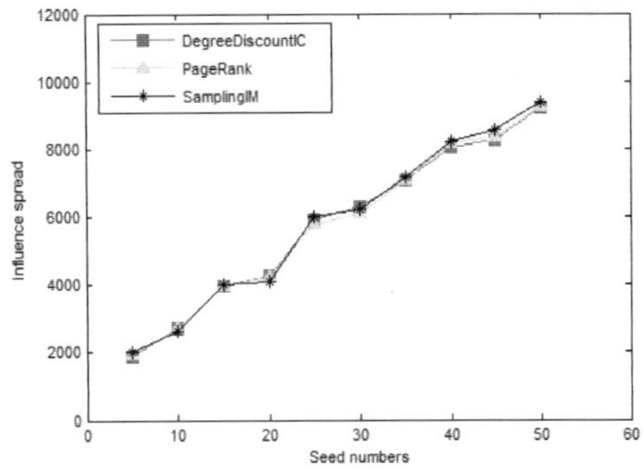

Figure 1:Comparison of the influence spreads of different algorithm on DBLP

From Figure 1, we can see that on DBLP datasets, the seed sets of different sizes selected by algorithm Sampling_IM can achieve the largest propagation range compared with algorithm PageRank and DegreeDiscountIC.

6.3.2. Result on Epinions. In the experiments on Epinions, we set $\varepsilon = \frac{1}{1000}$, $\rho = 6$, $\delta = 0.99$ By (10) we get $r = 351600$, and we sampled 351600 possible worlds in Epinions in our experiment. We compare the influence spreads of different algorithms, and the result is shown in Figure 2.

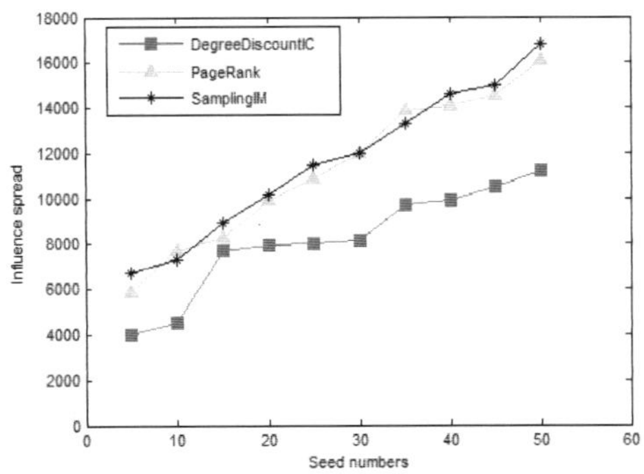

Figure 2:Comparison of the influence spreads of different algorithm on Epinions

From Figure 2, we can see that on Epinions datasets, the seed sets of different sizes selected by algorithm Sampling_IM can achieve the largest propagation range compared with DegreeDiscountIC. Compared with PageRank, Sampling_IM can achieve larger propagation range in most situations. And with expansion of the seed set size, Sampling_IM can achieve larger propagation range increment than PageRank.

6.3.3. Results on NetHPET. if In the experiments on NetHPET, we set $\varepsilon = \frac{1}{1000}$, $\rho = 6, \delta = 0.99$ Using (10), we get $r = 351600$, therefore we sampled 351600 possible worlds in the experiment on DBLP. Comparison of the influence spreads of different algorithms is shown in Figure 3.

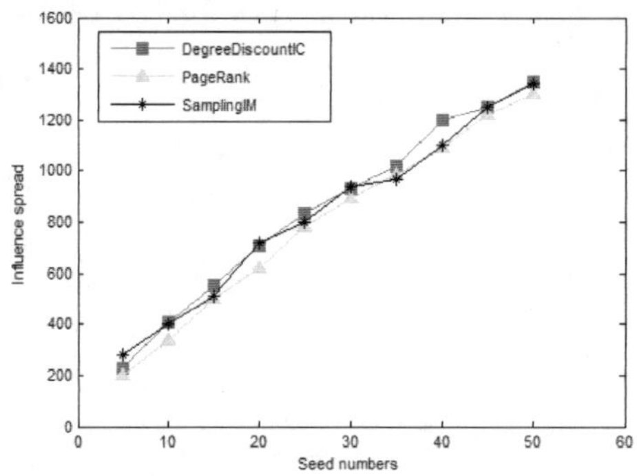

Figure 3: Comparison of the influence spreads of different algorithm on NetHPET

From Figure 3, we can see that on NetHPET datasets, the seed sets of different sizes selected by algorithm Sampling_IM can achieve almost the same propagation range compared with algorithm PageRank and DegreeDiscountIC. In some cases, the spread of Sampling_Imis is larger than the PageRank algorithm.

The experimental results above show that the proposed algorithm Sampling_IM can select seed set with larger propagation influence compared with the other two algorithms. Therefore, the algorithm Sampling_IM outperforms DegreeDiscountIC and PageRank in terms of influence spread. Furthermore, since Sampling_IM uses the sampling method instead of enumerating all the possible worlds in the network, it can greatly reduce the computation time.

7. Conslusions

In social networks, how to select one or more users that will make maximum influence on other users is a critical problem. In this paper, we present a sampling based algorithm for influence maximization on linear threshold model. The method uses Chernoff bound to estimate the size of the sampling possible worlds so that the error can be constrained within a given bound. Then the active possibilities of the routes in the possible worlds are calculated, and are used to compute the influence spread of each node in the network. The nodes with the largest influence spread are selected to form the seed set. Our experimental results show that our method can effectively select appropriate seed nodes set that can spread larger influence than other similar methods, and we use sampling technique to improve the efficiency.

8. References

[1] D.Kemp, J.M. Kleinberg, E. Tardos 2003 Maximizing the spread in influence through a social network, in KDD
[2] J. Leskovec et al. In KDD 2007 Cost-effective outbreak detection in networks.
[3] A. Goyal, W. Lu, and L. V. S. Lakshmanan, 2011 , "CELF++: Optimizing the greedy algorithm for influence maximization in social networks," in WWW (Companion Volume).
[4] W. Chen, Y. Wang, and S. Yang, in KDD 2009 "Efficient influence maximization in social networks,".
[5] Chen W, Wang C, Wang Y. in KDD 2010 "Scalable influence maximization for prevalent viral marketing in large-scale social networks,".
[6] Pavel Berkhin. 2005, 2(1):73-120 A Survey on PageRank Computing[J]. Internet Mathematics.
[7] Kim J, Kim S K, Yu H. 2013 Scalable and parallelizable processing of influence maximization for large-scale social networks. In ICDE, pages 266-277.

Acknowledgments

This research was supported in part by the Chinese National Natural Science Foundation under grant No. 61379066, 61070047, 61379064，61472344, 61402395, 61602202, Natural Science Foundation of Jiangsu Province under contracts BK20130452, BK2012672，BK2012128, BK20140492 and Six talent peaks project in Jiangsu Province(Grant No. 2011-DZXX-032).

IOP Conf. Series: Journal of Physics: Conf. Series **989** (2018) 012014 doi:10.1088/1742-6596/989/1/012014

Measurements of the Received Signal Level and Service Coverage Area at the IEEE 802.11 Access Point in the Building

N Gunantara[1], P K Sudiarta[1], AAN A I Prasetya[1], A Dharma[1], and I N Gde Antara[2]

1 Electrical Engineering, Universitas Udayana
2 Mechanical Engineering, Universitas Udayana
E-mail: gunantara@unud.ac.id

Abstract. Access point (AP) is part of a Wireless Local Access Network (WLAN) with its communications using WiFi. AP is used to transmit and receive data to users/clients. The ability of AP to serve users/clients depends on many factors. Moreover, if AP is applied in conditions inside the building. In this study, AP is installed at two points inside the building and then measured in the form of the received signal level (RSL) and service coverage area. One AP measured its performance by 26 measurement points and the other AP measured its performance by 20 measurement points. When AP has measured its performance then another AP position is switched off. Based on the measurement result, the received signal level value is the highest value is about -47 dBm at a distance of 3.2 m, while the lowest is about -79 dBm at a 9.21 m because it is on barrier 2 walls. While based on service coverage area, the area which is far away from the AP then the quality of service becomes bad because the transmitted signal is weakening caused by the distance and the loss of the wall.

1. Introduction

Access point (AP) is part of local radio access network where communications using WiFi. AP is tasked to transmit and receive data to users / clients. The existence of AP is very eagerly awaited by the community because with the AP we can communicate / access the internet cheaply and flexible than cable technology. The quality of service from the AP depends not only on the distance of the user with the AP only. Users who are close to the AP are not guaranteed to have good quality and vice versa for longer distance is not necessarily the worse. This is caused by many factors. Especially for conditions inside the building. Measurable performance in AP systems is received signal, throughput, channel capacity, and service coverage area.

AP placement is very influential on the performance of service area coverage. If not properly planned it will cause interference between the scope of service from the AP. Measurements were made using ORBIT testbed. Performance measured is performance against installed AP density. Interference across the scope of services from APs can be avoided with well-planned AP placements, frequency selection, and setting of AP parameters [1]. In addition, performance analysis of the effect of AP density on Finite-Area Networks was performed [2].

For AP placement in the building, received signal strength (RSS) is affected by multiple paths of AP signals, signal attenuation, and changes in environmental conditions within the building. To overcome this condition, this research is used with k-nearest neighbor (KNN) algorithm for location accuracy in building. The result obtained is improved location accuracy and performance from AP to be better [3]. Accuracy of WiFi localization is measured through a strong collection of signals received within the building by the WiFi trilateration method [4].

In this study, AP is installed at two points within a building and then measured receive signal level and service coverage area. When one AP is measured then the other AP's position is switched off, then receive signal level and service coverage area is measured. The receive signal level is measured by WiFi Analyzer while the service coverage area is measured by the Heatmapper Ekahau. The results of this measurement aims to improve the ability or performance of existing AP systems.

2. Wireless Local Area Network (WLAN)

The WLAN architecture consists of two services: a basic service set or also called a Basic Service Set (BSS) and an Advanced Service Set (ESS). The minimal BSS configuration consists of one access point and several users. The Extended Service Set (ESS) is a network established to provide extended coverage for wireless or wired users. Extended Service Set consists of several Basic Service Set (BSS) that overlap each other. The connection between two or more BSS infrastructure generates ESS. The architecture of the WLAN can be seen in figure 1 below [5]:

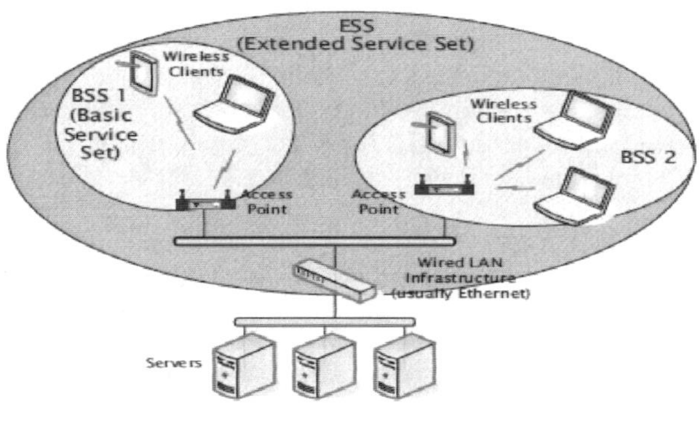

Figure. 1 Arsitektur WLAN

To build a WLAN network configuration, there are several main components required for WLAN networks to connect: Access Point, Wireless LAN Interface, and Mobile / Desktop PC [6]-[7]. AP used in this research is from Unifi product [8].

2.1. Indoor Propagation

In wireless propagation, the success of the information signal to destination will be affected by the surrounding environment. The information signal experiences reflection, absorption, distortion, or obstruction. To understand the propagation inside the building, it is necessary to know the propagation characteristics between the floor and the floor in a building. The propagation path between floors within a building must be determined. The amount of wall attenuation depends on the material used from the wall and the frequency of the passing signal [9]. Thus the received power of indoor propagation which is affected by the wall attenuation becomes [10]:

IOP Conf. Series: Journal of Physics: Conf. Series **989** (2018) 012014 doi:10.1088/1742-6596/989/1/012014

$$P_r = P_t G_t G_r \left(\frac{\lambda}{4\pi}\right)^2 d^{-\alpha} 10^{\frac{X_\varphi}{10}} \prod_{m=1}^{M} |\Gamma_m|^2 \quad (1)$$

where P_t is the transmitting power of the AP, G_t is the gain of the transmitting antenna, G_r is the gain of the receiving antenna, d is the AP distance and the user, α is power of distance, φ is the standard deviation of the normal log shadowing, Γ_m and M respectively denotes the transmission coefficient of the *m*-th wall is passed by the direct propagation path and the number of walls.

2.2. Performance of Access Point

Performance measured in this research is Receive Signal Level and Coverage service area.

2.2.1. Receive Signal Level (RSL)

RSL is an indication of the received power level of an antenna. Therefore, the higher the number of RSLs indicates that the stronger the signal is obtained. Through mathematical calculations that the receive signal can be calculated using equation (1). The obtained signal power is in dBm. While through measurement, received signal power is measured using Wifi Analyzer. Wifi Analyzer can be used to analyze the surrounding Wifi network. This app can get signal quality and network saturation information in just 5 seconds.

2.2.2. Service area coverage

A service area coverage of an AP is required to know the range of installed AP signals and to expand the network for optimal service delivery. Ekahau Heatmapper is software for mapping service area coverage of Wi-Fi networks. This software is easy to use to show the range of installed wireless AP network by adding digital maps.

This software is widely used to observe the coverage area in a place like home, office, school and others. Thus, users easily know where the place is not covered wireless network Wi-Fi. Ekahau Heatmapper software does not provide digital maps, so users are expected to have a digital map. The survey result of Ekahau Heatmapper software, where there are several different colors on each distance difference from AP and user. These colors indicate signal strength. The color differences shown in table 1 [11]:

Table 1. The Colorful Heatmaps Indicate Signal Strength

Colorful Heatmaps	Signal Strength (dBm)	Signal Quality Category
	-40 up to -50	Excellent
	-50 up to -60	Good
	-60 up to -85	Less
	-85 up to -100	Poor

3. Measurement Method

3.1. Room Plan

RSL measurements and service area coverage were carried out in indoor environments of buildings in Dishubkominfo Kabupaten Badung. Room plan for measurement can be seen in figure 2. From figure 2 it can be seen that there are 2 AP mounted on the hall of the building. The first AP was installed on the north and the second AP was installed on the south side of the building. Users are located in each room of the building.

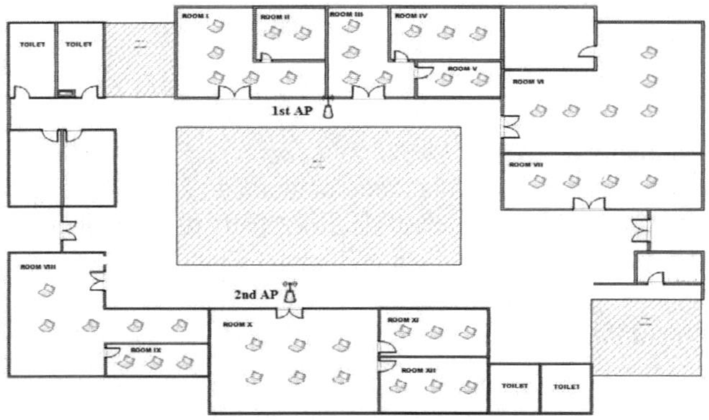

Figure 2. Plan Room Measurement

3.2. Measurement of Received Signal level

RSL measurements were performed for the condition of a wall barrier between the AP and the user, where RSL was measured according to the distance between the AP and the user. Measurements are made using an AP device called Ubiquiti wifi UAP and a laptop device that acts as a server. The function of the laptop server to set the Tx rate used. In addition to hardware devices, to perform RSL measurements required software installed on the smartphone client. The software is called Wifi Analyzer. In the software, the measurement is done by making sure the AP is connected to the smartphone by locking one of the AP signals you want to measure. Then run the software so that it will automatically look the measurement results in the form of Received Signal level.

3.3. Measurement of Service Coverage Area

Measurement of service coverage area using Ekahau HeatMapper software. It was done with the aim to know service coverage area in indoor environment at Dishubkominfo Badung, Bali, Indonesia. Further measurement results are analyzed to provide recommendations that the need for additional access points to cover the entire indoor area.

The service coverage area measurement is based on the received signal value previously measured at the measurement points. The quality of the service coverage area of the heatmapper ekahau software depends on the number of wall barriers passed by the AP signal.

4. Measurement Results

4.1. RSL For AP 1

RSL measurement results for AP 1 for 26 points spread over seven rooms. Users in the room, some are blocked by one and two walls. The results of these measurements are then sorted by the closest distance to the furthest distance from the AP position. The results can be seen in table 2 and shown in the graph shown in figure 3.

Table 2. RSL of AP 1

No.	Distance (m)	RSL (dB)	No.	Distance (m)	RSL (dB)
1.	3.2	-47	14.	8.06	-77
2.	3.77	-55	15.	8.07	-61
3.	3.9	-55	16.	8.6	-62
4.	4.24	-55	17.	9.21	-79

5.	5.2	-57	18.	9.5	-62
6.	5.4	-58	19.	9.89	-64
7.	5.83	-58	20.	9.96	-65
8.	7.2	-59	21.	11.5	-64
9.	7.28	-72	22.	11.88	-66
10.	7.28	-73	23.	13.5	-70
11.	7.5	-60	24.	13.64	-70
12.	7.61	-61	25.	13.82	-72
13.	8.06	-77	26.	14.08	-76

From figure 3 it shows that RSL at some point of measurement decreased significantly. Maximum RSL obtained is -40 dBm and lowest RSL reach -79 dBm. The decrease in RSL values occurs with increasing user distance to the AP. Based on equation (1), the RSL value depends on distance, shadowing, and damping of the wall. There are several points with almost the same distance having different RSL values. This is because the user with almost the same distance is located in a different room where one user is blocked by one wall and another user is blocked by two walls.

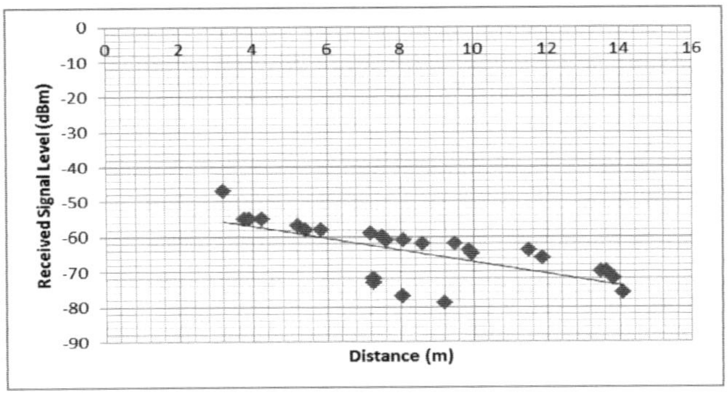

Figure 3. RSL and Distance for AP 1

4.2. RSL For AP 2

RSL measurement results for AP 2 for 2 points spread over five rooms. Users in the room, some are blocked by one and two walls. The results of these measurements are then sorted by the closest distance to the furthest distance from the AP position. The results can be seen in table 3 and shown in the graph shown in figure 4.

Table 3. RSL of AP 2

No.	Distance (m)	RSL (dB)	No.	Distance (m)	RSL (dB)
1.	4.5	-50	11.	8.9	-67
2.	4.92	-52	12.	9.43	-69
3.	4.92	-54	13.	9.48	-58
4.	6.26	-55	14.	9.96	-58
5.	6.5	-56	15.	10.25	-73
6.	6.8	-56	16.	11.18	-74
7.	6.8	-56	17.	11.4	-62

8.	7.61	-56	18.	11.8	-78
9.	7.81	-67	19.	13.03	-65
10.	8.07	-58	20.	13.34	-66

From figure 4 it shows that RSL at some point of measurement decreased significantly. Maximum RSL obtained is -50 dBm and lowest RSL reach -78 dBm. The decrease in RSL values occurs with increasing user distance to the AP. Based on equation (1), the RSL value depends on distance, shadowing, and damping of the wall. There are several points with almost the same distance having different RSL values. This is because the user with almost the same distance is located in a different room where one user is blocked by one wall and another user is blocked by two walls.

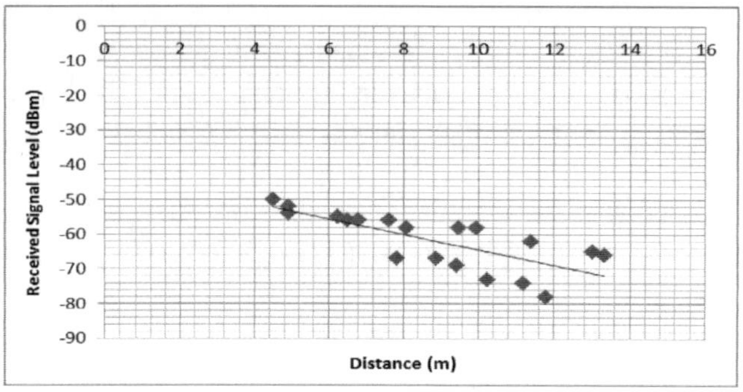

Figure 4. RSL and Distance for AP 2

4.3. Service Coverage Area for AP 1

Measurement of service coverage area for AP 1 is done by first switching off AP 2. The measurement is done based on the route of the users seen in figure 5. The result of Ekahau HeatMapper software can be seen in figure 5. From figure 5 it can be seen that the service quality of AP 1 shown in the form of colors. The dark green color indicates excellent service quality lies in areas close to AP 1. Light green means good service quality and yellow color indicates less service quality for users located further from AP 1. In addition to color, the Ekahau HeatMapper software also appears traces -the movement of users and hotspots that are active around at the time of measurement. The results of the measurements are then used as a consideration for the addition of AP to cover the entire room in the building.

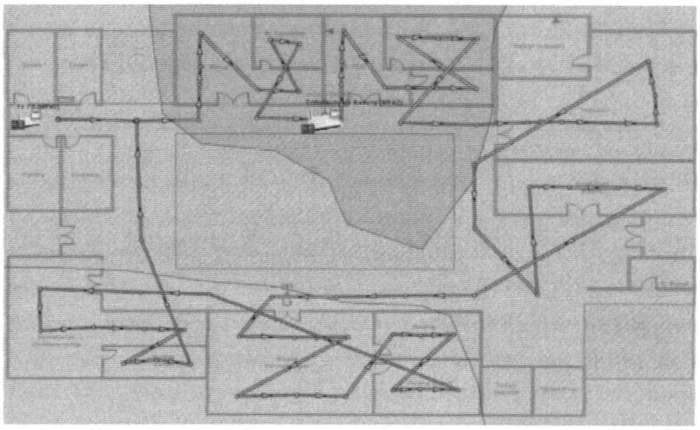

Figure 5. Service Coverage Area for AP 1

4.4. Service Coverage Area for AP 2

Measurement of service coverage area for AP 2 is done by first switching off AP 1. The measurement is done based on the route of the users seen in figure 6. The result of Ekahau HeatMapper software can be seen in figure 6. From figure 6 it can be seen that the service quality of AP 2 shown in the form of colors. The dark green color indicates excellent service quality lies in areas close to AP 2. Light green means good service quality and yellow color indicates less service quality for users located further from AP 2. In addition to color, the Ekahau HeatMapper software also appears traces -the movement of users and hotspots that are active around at the time of measurement. In contrast to the measurement results in AP 1, on the measurement of AP 2 detected many hotspots are active. The results of the measurements are then used as a consideration for the addition of AP to cover the entire room in the building

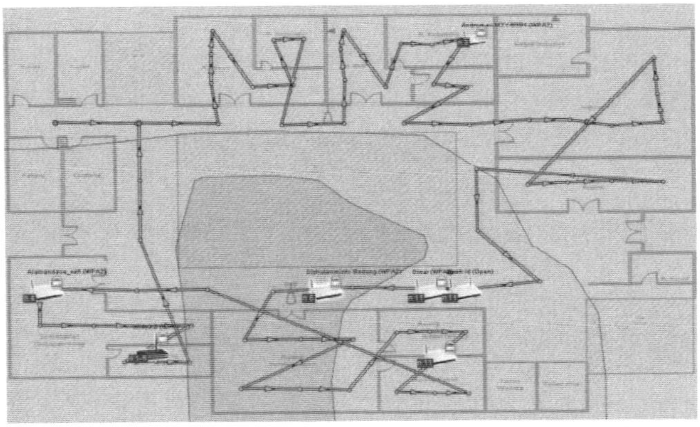

Figure 6. Service Coverage Area for AP 2

5. Conclusions

AP network performance measured is RSL and Service Coverage Area by using Wifi Analyzer and Ekahau HeatMapper. AP performance measurements were performed in buildings in Dishubkominfo. From the measurement results obtained the performance of AP 1 is RSL done on 26 points spread in seven rooms. RSL at several points of measurement decreased significantly with the maximum RSL obtained is -40 dBm and lowest RSL reach -79 dBm. The performance of AP 2 is done for 20 spots spread over five rooms. RSL obtained at some measurement points decreased significantly. Maximum RSL obtained is -50 dBm and lowest RSL reach -78 dBm. The decrease in RSL values occurs with increasing user distance and AP. Based on equation (1), the RSL value depends on distance, shadowing, and damping of the wall. The performance of service coverage area, service quality of AP is shown in the form of colors. The dark green color indicates excellent service quality, light green means good service quality, and yellow shows a less service quality.

Reference

[1] Ergin M A, Ramachandran K, and Gruteser M, **2007** Understanding the Effect of Access Point Density on Wireless LAN Performance Proceedings of the 13th annual ACM international conference on Mobile computing and networking

[2] Alireza Banani S., Andrew W.E., and Raviraj S. A., **2015** Analyzing the Impact of Access Point Density on the Performance of Finite-Area Networks IEEE Transactions on Communications vol. 63, issue 12 Dec 2015

IOP Conf. Series: Journal of Physics: Conf. Series **989** (2018) 012014 doi:10.1088/1742-6596/989/1/012014

[3] Yu F, Jiang M, Liang J, Qin X, Hu M, Peng T, and Hu X, **2014** 5G WiFi Signal-Based Indoor Localization System Using Cluster k-Nearest Neighbor Algorithm International Journal of Distributed Sensor Networks Volume 2014
[4] Bobescu B and Alexandru M, **2015** Mobile Indoor Positioning Using WiFi Localization, Review of the Air Force Academy vol. 28, no. 1
[5] Gultom, D.I. **2011**. Packet Delay Analysis Voip On Ad-Hoc Wireless Lan Network (IEEE 802.11). Universitas Sumatera Utara [Institutional Repository].
[6] Alexander T, **2007**, Optimizing and Testing WLANs Proven Techniques for Maximum Performance, Elsevier Inc : United States of America
[7] Razak, I dan Ulfiah, F. **2009**. A Study Of The Signal Quality Characteristics Of Building Profiles With Radio Propagation Modeling In The Indoor WLAN System, Media Elektrik, vol. 4, no. 1 [Indonesian Proceedings]
[8] https://www.ubnt.com
[9] Rappaport, T. S., **2002**, Wireless Communication Principles and Practice, Prentice-Hall, USA.
[10] Gunantara N and Hendrantoro G, **2013** "Multi-Objective Cross Layer Optimization for Selection of Cooperative Path Pairs in Multihop Wireless Ad hoc Networks", Journal of Communications Software and Systems, vol. 9, no. 3.
[11] https://www.ekahau.com/products/heatmapper/faq/#pane3

IOP Conf. Series: Journal of Physics: Conf. Series **989** (2018) 012015 doi:10.1088/1742-6596/989/1/012015

Flood Scenario Simulation and Disaster Estimation of Ba-Ma Creek Watershed in Nantou County, Taiwan

S H Peng[1] and Y K Hsu[1]

[1] Department of Spatial Design, Chienkuo Technology University, Changhua City, Taiwan, R.O.C.
E-mail: librakyo@cc.ctu.edu.tw

Abstract. The present study proposed several scenario simulations of flood disaster according to the historical flood event and planning requirement in Ba-Ma Creek Watershed located in Nantou County, Taiwan. The simulations were made using the FLO-2D model, a numerical model which can compute the velocity and depth of flood on a two-dimensional terrain. Meanwhile, the calculated data were utilized to estimate the possible damage incurred by the flood disaster. The results thus obtained can serve as references for disaster prevention. Moreover, the simulated results could be employed for flood disaster estimation using the method suggested by the Water Resources Agency of Taiwan. Finally, the conclusions and perspectives are presented.

1. Introduction

In recent years, increasing and frequent hazards caused by typhoons and heavy rainfalls in Taiwan have resulted in huge damages and financial loss. Local drainage in urban and rural regions plays an important role in flood control [1]. For this reason, the project known as "Outline of Flood Control in Potential Flood Districts" has been implemented in Taiwan. The aim of the project is to decrease the inundation disasters caused by typhoons and torrential rainfalls in regions located in proximity to minor streams or local drainage.

After the Chi-Chi earthquake in 1999, the large amount of loose debris has resulted in floods or landslides under heavy rainfall in Nantou County, Taiwan. In 2001, the heavy downpour brought by Typhoon Toraji triggered the overflow of embankment in Ba-Ma Creek, causing the serious disasters in many villages. According to the local residents, Typhoons Herb, Toraji and Mindulle had flooded a total area of about 36 hectares. Hence, estimation of flooding range or damage using numerical models of flood inundation has become very important.

The models of flood inundation have different theoretical background and assumptions. One-dimensional steady flow [1] or unsteady flow [2] model is relatively simple, less initial conditions and calculation time. Two-dimensional model [3, 4] can simulate the movement of flood inundation zone, but the calculation requires more information to construct the computed domain, and longer computer algorithm time to analyze simulation in flooded terrain. Besides, many previous studies [5-7] also provided very valuable reference.

This study simulated several scenarios of flood inundation using the FLO-2D model [8]. With the simulated results, flood damages of various return periods were estimated, which could serve as references for planning flood control strategy.

2. Materials and methods

2.1. Site description

The main region of Ba-Ma Creek Watershed, with an area of 709.1 ha, is located in Shui-Li Township, Nantou County, Taiwan. The main stream and two branches are 6.475 km, 1.11 km and 1.878 km long, respectively. Manning's roughness coefficient for scenario simulation was obtained by investigation of riverbed particle size [9]. A 10-year or 25-year flood frequency was adopted as the criterion for peak flow calculation in order to plan or design the capacity of channels. The current design standard is a 10-year return period including an extra height of 0.5 m over the bank, and the water level should not overflow the levees for a 25-year return period flood at the same time.

Many typhoon and storm events occurred during 1957 to 2015 were used in the hydrological analysis. The peak flow discharges for various return periods at different hydrological stations were also calculated based on the design storm for 1-day using the rational method, triangle unit hydrograph method and instantaneous unit hydrograph, respectively [10]. Finally, the peak flow discharges estimated by the triangle unit hydrograph method for 2-, 5-, 10-, 25- and 50-yr return periods are suggested to be the discharges for simulations.

2.2. Numerical model

Hydraulic computations involving 1D and 2D modeling were also carried out. In 1D simulation, the Hydraulic Engineering Center: River Analysis System (HEC-RAS) model was employed to compute the flood level. The results thus obtained can assist in planning the locations and heights of levees and checking the stream cross-sectional area of existing structures such as bridges and culverts. 2D simulations were also calculated using the FLO-2D model, a two-dimensional flood routing model used by engineers and floodplain managers to predict flood hazards, and flood damages can be evaluated from the simulation results [11]. The general constitutive fluid equations including the continuity equation and the equation of motion (dynamic wave momentum equation) [3] are

$$\frac{\partial h}{\partial t}+\frac{\partial (hu)}{\partial x}+\frac{\partial (hv)}{\partial y}=i \quad (1)$$

$$S_{fx}=S_{ox}-\frac{\partial h}{\partial x}-\frac{\partial u}{g\partial t}-u\frac{\partial u}{g\partial x}-v\frac{\partial u}{g\partial y} \quad (2)$$

$$S_{fy}=S_{oy}-\frac{\partial h}{\partial y}-\frac{\partial v}{g\partial t}-u\frac{\partial v}{g\partial x}-v\frac{\partial v}{g\partial y} \quad (3)$$

where h is the flow depth (m); u, v are the depth-averaged velocities in the x- and y-axis directions (m/s); i is the excess rainfall intensity (mm/hr); S_{fx}, S_{fy} are the friction slope components according to Manning's equation of the x- and y-axis directions; S_{bx}, S_{by} are the bed slopes of the x- and y-axis directions; and g is acceleration of gravity (m/s^2).

3. Results

3.1. Scenario simulation

To compare the simulated and real flood inundations, rainfall data of Typhoon Toraji in 2001 were employed to compute the flood depths using the FLO-2D model. Meanwhile, site visits were made to understand the flood range caused by Typhoon Toraji and to survey the flood mark for describing the inundated regions, which covered approximately 39.2 ha. The maximum depth was around 3 m along the bank of the main stream near the National Shui-Li Vocational High School of Industry and Commerce. In this scenario simulation, the numerical result indicated that the main flood area was about 19.2 ha and the maximum depth along the main stream bank near the school was approximately 4 m (see figure 1 and table 1).

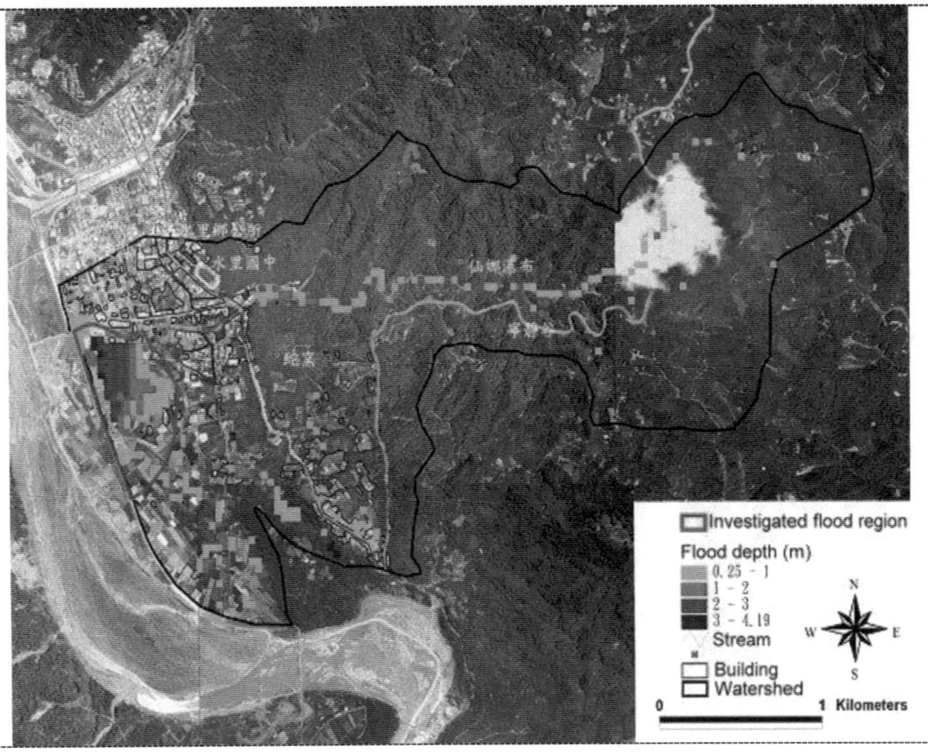

Figure 1. Comparison of simulated and investigated flood regions.

Table 1. Areas and depths of scenario simulation of Typhoon Toraji

Average depth (m)	Maximum depth (m)	Flood area (ha)				Total flood area (ha)
		Depth 0.25-1 m	Depth 1-2 m	Depth 2-3 m	Depth >3 m	
1.092	4.19	31.04	8.8	6.72	0.96	47.52

3.2. Numerical model

In addition to the scenario simulation of Typhoon Toraji, simulations of 2-, 5-, 10-, 25- and 50-yr return periods were also completed, respectively. With these simulated flood areas and depths, the flood damage could be estimated. First, the flood areas and depths for various land use types were extracted from the simulated results using the GIS software ArcGIS. Then the damage for farms, buildings and other land uses at various depths were estimated using the method suggested by the Water Resources Agency of Taiwan [12]. Due to the lack of statistical data of flood disaster in Ba-Ma Creek Watershed, therefore, this study estimated the flood damage based on the scenario simulation results and economic situations. The annual average flood losses could be considered from agricultural losses, building losses, public facilities and other losses. The direct agricultural losses were estimated by the statistical analysis of farmland flooding depth and crop yield reduction curve. Building losses were estimated in three parts, namely housing, household goods and business units (commercial, factory), and then calculated the total value of flood losses according to the affected area. Public facilities and other losses were presumed to be based on estimates of past flood damage statistics or the rates of loss of public facilities floods and general assets floods. From this analysis (Table 2), the damages of 2-, 5-, 10-, 25- and 50-yr return periods can be realized. The engineers can decide how much should be invested in the engineering works of flood control or environmental construction plans.

4. Conclusions

In this study, flood inundations for various return periods were simulated and damages were estimated using the method suggested by the Water Resources Agency of Taiwan. There is a little difference between the scenario simulation and real situation in the application of numerical model because the assumptions and simulation parameters are moderately simplified. However, these results of the simulation still offer a considerable benefit for the qualitative trend, and more detailed information such as water depth and range. Although these data are for reference only and do not totally represent the actual situations, at least they may provide the quantifiable data to reflect the relative extent of the different conditions to each other. According to the damage analysis, the total preliminary estimated cost for the construction and repair of the hydraulic structures is around NT$ 107,480,000. The implementation of this planning will be helpful for channel stabilization and prevention of flood disaster. Furthermore, the flood control project is also useful for the river restoration or planning of environmental construction if the engineering works are finished.

Table 2. Flood damage estimations for 2-, 5-, 10-, 25- and 50-yr return periods

Return period (yr)	Depth (m)	Farm land		Building land		Other land		Total damage (NT$ 1,000)
		Flood area (ha)	Damage (NT$ 1,000)	Flood area (ha)	Damage (NT$ 1,000)	Flood area (ha)	Damage (NT$ 1,000)	
2	0.25-1	8.09	469.2	1.35	3,375.0	4.89	283.6	8,599.4
	1-2	5.73	630.3	0.67	3,350.0	0.34	54.4	
	2-3	0.51	76.5	0.00	0.0	0.17	81.6	
	>3	0.17	34.0	0.00	0.0	0.17	244.8	
5	0.25-1	9.77	566.7	1.69	4,225.0	6.07	352.1	10,109.4
	1-2	7.92	871.2	0.67	3,350.0	0.51	81.6	
	2-3	1.35	202.5	0.00	0.0	0.17	81.6	
	>3	0.67	134.0	0.00	0.0	0.17	244.8	
10	0.25-1	11.63	674.5	1.52	3,800.0	6.57	381.1	10,985.9
	1-2	6.40	704.0	0.84	4,200.0	0.84	134.4	
	2-3	4.21	631.5	0.00	0.0	0.17	81.6	
	>3	0.67	134.0	0.00	0.0	0.17	244.8	
25	0.25-1	13.65	791.7	1.69	4,225.0	7.58	439.6	11,999.5
	1-2	5.22	574.2	0.84	4,200.0	0.84	134.4	
	2-3	6.74	1,011.0	0.00	0.0	0.00	0.0	
	>3	0.67	134.0	0.00	0.0	0.34	489.6	
50	0.25-1	14.49	840.4	1.69	4,225.0	7.58	439.6	12,233.6
	1-2	5.39	592.9	0.84	4,200.0	1.35	216.0	
	2-3	7.08	1,062.0	0.00	0.0	0.00	0.0	
	>3	0.84	168.0	0.00	0.0	0.34	489.6	

5. References

[1] Brunner G W 2006 HEC-RAS river analysis system – hydraulic reference manual. Davis: USACE Publishing, pp. 1-7

[2] Sarveram H, Shamsai A and Banihashemi M A 2012 One-dimensional dam-break flow modeling for variable width channels using an implicit semi-Lagrangian method. Journal of Food, Agriculture & Environment **10**(1): 835-838

[3] O'Brien J S 2003 FLO-2D users manual. Nutrioso: FLO Engineering Inc. Publishing, pp. 7-11.

[4] Yin J, Yu D, Yin Z, Wang J and Xu S 2013 Multiple scenario analyses of Huangpu River flooding using a 1D/2D coupled flood inundation model. Nat Hazards **66**(2): 577–589

[5] Yin J, Ye M, Yin Z and Xu S 2015 A review of advances in urban flood risk analysis over China. Stoch Environ Res Risk Assess **29**(3): 1063-1070

[6] Li W, Xu B and Wen J 2016 Scenario-based community flood risk assessment: a case study of Taining county town, Fujian province, China. Nat Hazards **82**(1): 193-208
[7] Leskens J G, Kehl C, Tutenel T, Kol T, Haan G, Stelling G and Eisemann E 2017 An interactive simulation and visualization tool for flood analysis usable for practitioners. Mitig Adapt Strateg Glob Change **22**(2): 307-324
[8] O'Brien J S, Julien P Y and Fullerton W T 1993 Two-dimensional water flood and mudflow simulation. J. Hydr. Engrg. ASCE **119**(2): 244-261
[9] Peng S H and Tang C 2013 Investigation of surface grain size in a mountainous gravel-bed creek in Taiwan using automated image analysis. Journal of Food, Agriculture & Environment **11**(2): 820-824
[10] Chow V T, Maidment D R and Mays L W 1988 Applied hydrology. Singapore: McGraw-Hill, pp. 459-465
[11] Peng S H and Lu S C 2013 FLO-2D simulation of mudflow caused by large landslide due to extremely heavy rainfall in southeastern Taiwan during Typhoon Morakot. J. Mt. Sci. **10**(2): 207–218
[12] WRPI 2006 Reference manual of river management and environmental planning. Water Resources Planning Institute, Water Resources Agency, Ministry of Economic Affairs, Taiwan: Taichung

Acknowledgments

This work was partly funded by the Nantou County Government, Taiwan. The assistance of Mr. Jian-Da Wu, P.E. of Soil and Water Conservation at Jian-Lei Engineering Co., Ltd. is gratefully acknowledged.

Layout Guide for Burnt and Un-burnt Tropical Forest: The Diversity of Forest Plants and Insetcs for Sustainable Environmental

N L Watiniasih[1], J Tambunan[1], I M Merdana[2] and I N G Antara[3]

[1] Department of Biology, Faculty of Mathematic and Basic Sciences, Udayana University, Bali, Indonesia
[2] Faculty of Animal Husbandry, Udayana University, Bali, Indonesia
[3] Department of Mechanical Engineering, Faculty of Engineering, Udayana University, Bali. Indonesia
E-mail: luhwatiniasih@unud.ac.id

Abstract. Forest fire is a common phenomenon in tropical forest likes in Indonesia. Beside the effect of soaring heat and lack of rain during dry season due to the tropical climate, farming system is also reported as one reason of forest fire in Indonesia. People of surrounding areas and neighbouring countries are suffering from the effect of forest fire. Plants and animals are the most suffer from this occurrence that they cannot escape. This study aimed to investigate the effect of previously burnt and un-burnt tropical forest in Borneo Island on the plant and insect diversity of the tropical forest. The result of the study found that the plants in previously burnt forest area was dominated by one species, while higher and more stable plant diversity was found in un-burnt forest. Although the number of individual insects was higher in previously burnt tropical forest, but the insects was more diverse in un-burnt tropical forest. The alteration of environmental conditions in previously burnt and un-burnt forest indicate that the energy held in natural forest support higher number and more stable insects than previously burnt forest.

1. Introduction

Indonesia has been well known as a country with megabiodiversity, that supported by the tropical rain forest. Forest is a germ source that has many beneficial [1]. Ecologically, forest is a storehouse of biological diversity, habitat of high number of animal species. It prevents soil erosion, stabilize and fertilize the soil. Economically, forest is a source of carbon stock, water resources, provides a source of non-timber products, and has been quite famous for sustainable ecotourism. Ethnologically, forest and its diversity are source of herbal or traditional medicine [2].

Indonesian government has been dependent on the forest since long time. Part of the forest has been cleared out for plantation. The cleared forest was planted with rubber tree (Hevea brasiliensis) and palm oil plants (Elaeis sp). Forest clearance using firing technique was commonly reported, as it causes devastated forest fire. This action has been banned by the Indonesian government. However, during the dry season in Indonesia, many part of forest are still burning. It was reported due to natural causes or human activities. The natural cause of the forest fire can be due to the effect of soaring heat and lack of rain during dry season. This resulted in heavy smoke around the burning forest as well as to the neighboring countries [3].

Forest or Ecosystem Restoration (ER) has been proposed by the Indonesian government since 2003 [1]. Operationally, to comply and develop comprehensive with the regulation, study has to be conducted to understand the first state of the forest condition. This was suggested by Heriansyah [1] that to develop comprehensive regulation, study can be start by collecting a basic data such as inventory of plant and animal at a site in order to know its ecological value of each species in the ecosystem. Many studies have been conducted on diversity of plant and insect, but most of the data was collected form temperate area [4]. This study aimed to investigate the diversity of insect in forest that has previously burnt in comparison to the forest that is still intact or has never been burnt. There was no data available for in regard yet..

2. Materials and methods

Research was conducted at Palangkaraya, West Kalimantan Island (Borneo Island, Indonesia), from July to August 2016. Sites were divided into two: 'un-burnt' peat forest sites, which the site has never been on fire, located in Bukit Batu Natural Park (2°16'29.7"S - 113°59'59.5"E), and 'burnt' peat forest sites, which the site has previously been burnt, but at the time of study the plant has been growing, located at Kalampangan Community Forest (2°02'11.2"S - 113°46'26.2"E) adjacent to community farm. In each site samples were collected by beating and aerial sweeping techniques. The beating technique was conducted by choosing a plant in site, then a branch as long as 0.5m was beaten for 30 second. Three branches were beaten from each plant for 5 plants in each site. Aerial sweeping was conducted by swinging the insect net for 2 minutes every time. Samples then transported to the Lab for further identification and examination.

3. Result and discussion

The forest area, at both sites, was damp peat forest with the soil pH of 4.5 – 5.0. Thirty six known species plants have been described at the study site (around 2000m2). Most of them (25 species) was woody tall plant as commonly found in tropical rain forest, 11 plants were shrub and ferns. The diversity of plant at un-burnt forest site was higher than that of burnt forest site (Table 1). All 31 described plant species was found at un-burnt forest site compared to that of 6 at previously burnt forest site.

Table 1. Plant species found at both study sites (un-burnt and burnt forest sites)

No.	Plant spesies	Un-burnt Forest site	Burned forest site
1	Koompassia malaccensis	√	
2	Syzygium cuneatum	√	
3	Palaquium sp	√	
4	Syzygium spp	√	
5	Dyera lowii	√	
6	Shorea balangeran	√	
7	Calophyllum spp	√	
8	Lophopetalum javanicum	√	
9	Tetramerista glabra	√	
10	Diospyros spp	√	
11	Macaranga rhizinoides	√	
12	Myristica maxima	√	
13	Elaeocarpus malacensis	√	
14	Shorea leprosula	√	
15	Shorea johorensis	√	
16	Lithocarpus conocarpa	√	√
17	Alstonia pneumatophora	√	
18	Xylopia fusca	√	

19	Gonystylus bancanus	√	
20	Campnosperma auriculata	√	
21	Combretocarpus rotundatus	√	
22	Melastoma malabatrichum	√	√
23	Acacia auriculiformis	√	√
24	Acacia mangium	√	√
25	Stenochlaena sp	√	√
26	Mimosa pudica*	√	√
27	Uncaria sp*	√	
28	Arcangelisia flava*	√	
29	Vitex pinnata*	√	
30	Pandanus helicopus*	√	
31	Nephrolepis bisserata*	√	
32	Rhodomyrtus tomentosa*	√	
33	Ilex samosa*	√	
34	Calophyllum exceltum*	√	
35	Dicranopteris linearis*	√	
36	Baccaurea bracteate*	√	

The result show that the total number of insect collected was higher in un-burnt forest (258 individuals) compared to burnt forest (225 individuals). The order of Hymenoptera particularly the family of Formicidae (ants) contributed to the largest number of individual insets found, which 89 individuals at un-burnt forest site compared to 160 individuals at burnt forest site. This number may not be as high as it was predicted due to the sampling technique employed that most of the ant study using pitfall, litter sampling [5] or baiting [6] As this study concern of the diversity of all type of insects in the two sites, the pitfall sampling was not highly considered. High number of insect order was found only at un-burnt forest site such as Odonata and Coleoptera. Other insect order such Blattoedea was also found at un-burnt forest site, while Mantodea at burnt forest site (figure 1). Coleoptera was found in high number in un-burnt forest site as they are most of the species are herbivore [7].

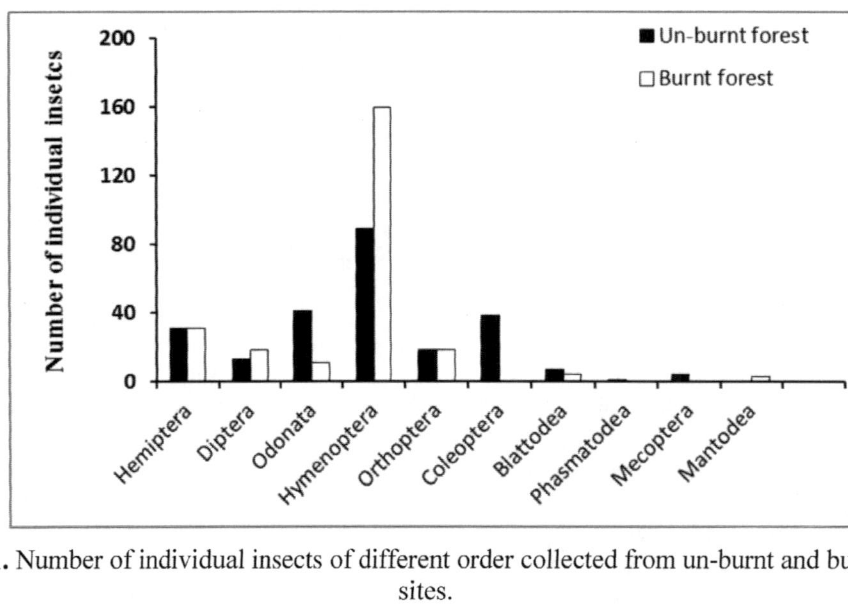

Figure 1. Number of individual insects of different order collected from un-burnt and burnt forest sites.

The high number family was from the order of Coleoptera (12 families) followed by Hemiptera and Hymnoptera (8 families each) (table 2). Many studied such as (Novotny, 2014 and Rinker 200) found out that apart from Coleoptera, Lepidoptera was commonly found in tropical forest. However, in this study there was no Lepidopteran found. There was not many flowering plants found during sample collection, therefore may not attract the Lepidopteran visiting the site. The hemipteran families were quite common found in the study sites, which also 4 families out of 8 families were found at burnt forest, namely the family of Coreidae, Miridae, Reduviidae and Cicadellidae. Reduviidae, the assassin bugs, was more commonly found in burnt forest site, three times compared to that of in un-burnt forest site.

Table 2. The number of individual insects collected from un-burnt and burnt forest sites in different order and families

Order	Family	Number of individual of un-burnt forest site	Number of individual of burnt forest site
Coleoptera	Cicindelidae	8	1
	Miridae	2	0
	Tessaratomidae	1	0
	Chrysomelidae	6	0
	Leiodidae	3	0
	Curculionidae	9	0
	Coccinellidae	2	0
	Anthicidae	1	0
	Tenebrionidae	2	0
	Cantharidae	3	0
	Cerambycidae	3	0
	Scirtidae	1	0
Hemiptera	Coreidae	5	1
	Alydidae	7	0
	Miridae	1	2
	Pentatomidae	1	0
	Reduviidae	7	21
	Cicadellidae	5	5
	Aphrophoridae	4	0
	Mesoveliidae	1	0
Hymenoptera	Vespidae	10	10
	Mutillidae	1	0
	Formicidae	75	136
	Ichneumonidae	2	1
	Sphecidae	1	0
	Argidae	1	0
	Braconidae	1	0
	Figitidae	11	0
Diptera	Culicidae	4	8
	Muscidae	5	10
	Drosophilidae	4	0
Odonata	Libellulidae	22	1
	Gomphidae	7	4
	Coenagrionidae	12	6

IOP Conf. Series: Journal of Physics: Conf. Series **989** (2018) 012016 doi:10.1088/1742-6596/989/1/012016

Orthoptera	Acrididae	12	3
	Gryllidae	5	4
	Tettigoniidae	1	5
Blattodea	Blattidae	3	0
	Blaberidae	4	4
Phasmatodea	Phasmatidae	1	0
Mecoptera	Panorpidae	4	0
Mantodea	Mantidae	0	3
Total number of individuals		**258**	**225**

The high diversity of insect collected in tropical forest was well known. Rinker and Lowman [7] states that the historical hypothesis focusing on long-term environmental stability may affect the higher present of species specialist in tropical area. There number of families present the forest that has never been burnt was three times higher compared to forest that has previously been burnt. One family only found at previously burnt forest site out of 12 families present during studies. Lower number of insect diversity was found in previously burnt sites may correspond with the lower number of plant diversity on this site. The higher temperature on the treetop stimulate foraging and ovipositioning of insects [8]. The abundance of young leaves provided at the un-burnt forest attracts more leaf feeding insect, in that younger leaves has less physical or chemical defense [7] It has been predicted that plant species richness can precisely predict and highly correlated to the insect species richness in tropical forest [9].

4. References

[1] Heriansyah I 2017 Forestry Research, Development and Innovation Agency (FORDIA) Ministry of Environment and Forestry Embassy of Republic of Indonesia, Tokyo
[2] Rates S M K *Toxicon* **39** 603
[3] Harrison M E, Page S E, Limin S H 2009 Biologist 56 156
[4] Nair K S S S 2007 Tropical Forest Insect Pest: Ecology, Impact and Managing (Cambridge University Press) p 393
[5] Ulysséa M A and Brandão C R F 2014 *Rev. Brasil. de Entomol.* **57** 217
[6] Dunn R R, Sanders N J, Fitzpatrick M C, Laurent E, LessardJ-P, Agosti D, Andersen A N, Bruhl C, Cerda X, Ellison A M, Fisher B L, Gibb H, Gotelli N J, Gove A, Guenard B, Janda M, Kaspari M, Longino J T, Majer J, McGlinn T P, Menke S B, Parr C L, Philpott S M, Pfeiffer M, Retana J, Suarez A V and Vas concelos H L 2007 Myrmecol. News 10 77
[7] Rinker H B and Lowman M D 2000 Insect Ecology: An Ecosystem Approach Insect Herbivory in Tropical Forests ed T D Schowalter
[8] Basset Y, Novotny V. Miller S E and Kitching R L 2003 Arthropods of Tropical Forests (Cambridge University Press) p 469
[9] Basset Y, Cizek L, Cuénoud P, Didham R K , Guilhaumon F, Missa O, Novotny V, Ødegaard F, Roslin T, Schmidl J, et al 2012 Arthropod diversity in a tropical forest Science, Am. Assoc. Advan. of Sci. 338 1481

Acknowledgments

Thanks to Institute of Research and Community Service, Udayana University for the encouragement that this manuscript can be made. The cooperation of the member of IMEET group was tremendous, so thank for that.

Determination Plastic Properties of a Material by Spherical Indentation Base on the Representative Stress Approach

I N Budiarsa[1], I N Gde Antara[2], Agus Dharma[3], I N Karnata[4]

[1] Mechanical Engineering University of Udayana, Indonesia
[2] Mechanical Engineering University of Udayana, Indonesia
[3] Electrical Engineering University of Udayana, Indonesia
[4] Civil Engineering University of Udayana, Indonesia
E-mail nyoman.budiarsa@unud.ac.id

Abstract. Under an indentation, the material undergoes a complex deformation. One of the most effective ways to analyse indentation has been the representative method. The concept coupled with finite element (FE) modelling has been used successfully in analysing sharp indenters. It is of great importance to extend this method to spherical indentation and associated hardness system. One particular case is the Rockwell B test, where the hardness is determined by two points on the P-h curve of a spherical indenter. In this case, an established link between materials parameters and P-h curves can naturally lead to direct hardness estimation from the materials parameters (e.g. yield stress (y) and work hardening coefficients (n)). This could provide a useful tool for both research and industrial applications. Two method to predict p-h curve in spherical indentation has been established. One is use method using C1-C2 polynomial equation approach and another one by depth approach. Both approach has been successfully. An effective method in representing the P-h curves using a normalized representative stress concept was established. The concept and methodology developed is used to predict hardness (HRB) values of materials through direct analysis and validated with experimental data on selected samples of steel.

1. Introduction

One significant advantage of indentation is that it only requires a small amount of materials; this makes it very attractive for the characterization of materials with gradient property where standard specimen is not readily available such as in situ or in vivo tests [1]. However, despite its wide use, the materials behaviour (represented by the hardness or P-h curves) are not explicitly linked with the constitutive material properties. Further work is required to be able to predict indentation resistance (P-h curves and/or hardness) from constitutive materials parameters. On the other side, it is also of great significance to both research and practical use to explore the potential to use indentation data to predict the constitutive materials properties. It may potentially provide a quicker way for material parameter identification and applicable in situation where standard specimen is not available. Developments of both areas require a detailed understanding/program linking constitutive materials properties, P-h curve and hardness from spherical indenters.

Earlier works showed that hardness can be related to the stress of the indented material, σr, corresponding to a representative strain, εr, which represents the mean plastic strain after yielding. The concept coupled with finite element (FE) modelling has been used successfully in analyzing sharp indenters where the representative strain and stress is well defined with a fixed indenter angle [2], [3]. It is of great importance to extend this to spherical indentation and associated hardness system. One

IOP Conf. Series: Journal of Physics: Conf. Series **989** (2018) 012017 doi:10.1088/1742-6596/989/1/012017

particular case is the Rockwell B test, where the hardens is determined by two points on the P-h curve of a spherical indenter. In this case, an established link between materials parameters and P-h curves can naturally lead to direct hardness estimation from the materials parameters (such as yield stress and work hardening coefficients). This could provide an useful tool for both research and practical applications.

2. Experimental

The material used were steel. The chemical compositions of materials are listed in table 1.

Table 1. Chemical compositions of the materials

Material	Condition	Element Composition (%)					
		C	Mn	P	S	Si,	Ni
Carbon Steel 0.10 % C	Norm.at 900°C	0.1	0.5	<0.04	<0.05	0.100	0.01
Mild Steel	N/A	0.3	0.3	0.05	0.05	0.122	490 ppm
Carbon Steel 0.54% C	Norm.at 840°C	0.54	0.9	0.055	0.014	0.19	0.014

Sample steel used is solid rod-shaped elliptical of 5 mm in diameter and 90 mm long. The tensile tests were performed with extensometer interfaced with a microcomputer, graphical print out of the test can be obtained and test data saved. The machine has a maximum loading capacity of 30 kN, with the readings being accurate to 0.5% of the force. The two main materials were used in this experiment include a carbon steel (0.1% C Steel) and Mild steel specimen. Typical The stress–strain curves of the two main materials tested are shown in figure 1a. Yield stress (σy) identified for the 0.1% C carbon steel is 308.03 MPa, work hardening coefficients (n) is 0.07; the yield stress (σy) for the mild steel is 601.66 MPa, the work hardening coefficient (n) is 0.025. The stress–strain curves and the material properties data are to be used as input to the FE model and to assess the accuracy of the P-h curves based hardness evaluation and inverse material properties. Specimens were sectioned, in order to hardness testing performed using hardness test equipment. The indentation using spherical indenter B scale, with R= 0.79 mm (Diameter steel ball =1/16 in), the indenter is forced into the sample under a preliminary minor load (Fm) = 10 Kgf and followed by a mayor loading (FM) =100. The hardness values is determined by the depth of penetration spherical indenter loaded on test samples following equation HR=130-(h1000-h100)/0.002. The machine is calibrated using a testing block, the variation of the hardness is within 5%. The average HRB values for two sample materials are shown in figure 1.b. The Rockwell hardness B (HRB) identified for the 0.1% C is 62.0, and HRB for Mild Steel is 88.0 These HRB hardness data be used to validate the representative stress based hardness evaluation and property prediction program to be developed.

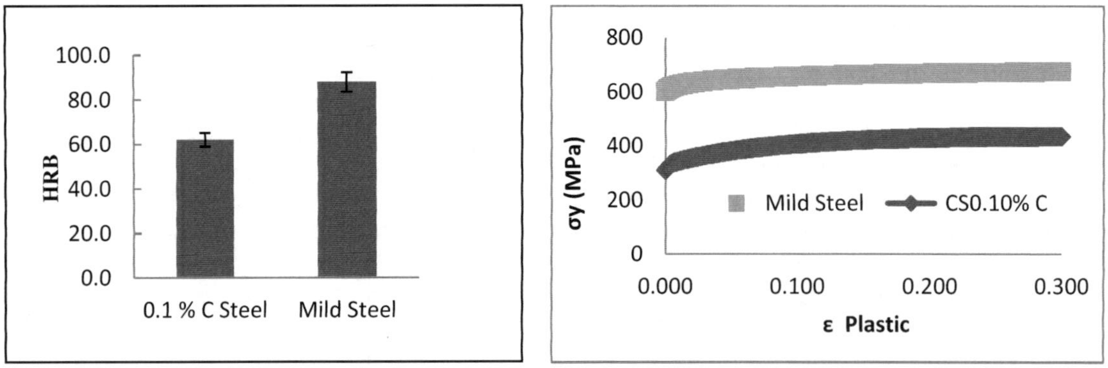

Figure 1. Typical Plastic stress strain data of the specimen 0.1% Carbon Steel and Mild Steel (2b) Rockwell (B scale) hardness data of the specimen (minor load=100N, Maximum load =1000).

IOP Conf. Series: Journal of Physics: Conf. Series **989** (2018) 012017 doi:10.1088/1742-6596/989/1/012017

3. Framework and analysis

3.1. Material model

The power law description, almost used to approximate the plastic behaviour of metal material [4]. A comprehensive framework using power law was developed earlier [5] in material model with reduce young modulus E*, within a specified range of material parameters has been identified a representative plastic strain and used to normalize the loading curvature of material hardening exponent (n). This work is concerned with elastic-plastic materials with a particular focus of steel, the stress-strain curves can be represented by Hooke's law and Von Misses yield criterion with isotropic power law hardening. Under those general conditions, the dependence of the true stress σ on the true strain e is commonly expressed:

$$\sigma = \begin{cases} \mathrm{E}\,\varepsilon & \text{for } \sigma \le \sigma_y \\ \sigma_y \left(\frac{\mathrm{E}}{\sigma_y}\right)^n \varepsilon^n & \text{for } \sigma > \sigma_y \end{cases} \tag{1}$$

Where E is the Young' s modulus, n the strain hardening exponent and σy the initial yield stress at zero offset strain. If er is a particular plastic strain point, the stress at the point representative stress, sr. At this plastic strain point, the stress can also be expressed as:

$$\sigma_r = \sigma_y \left(1 + \frac{E}{\sigma_y}\,\varepsilon_r\right)^n \tag{2}$$

3.2. Numerical model

Axial symmetric 2-D space FE models were constructed to simulate the indentation response of elastic plastic solids, are shown in figure 2. Model was used due to the symmetry of the spherical indenter. The indenter was assumed to be rigid body as it is much harder than the indented material. The movement of the indenter was simulated by displacing a rigid arc (rigid body) along the Z axis. In the model, the sample size can be changed to ensure that the sample is much larger than the indenter radius/contact area during the indentation to avoid potential sample size and boundary effects [6]

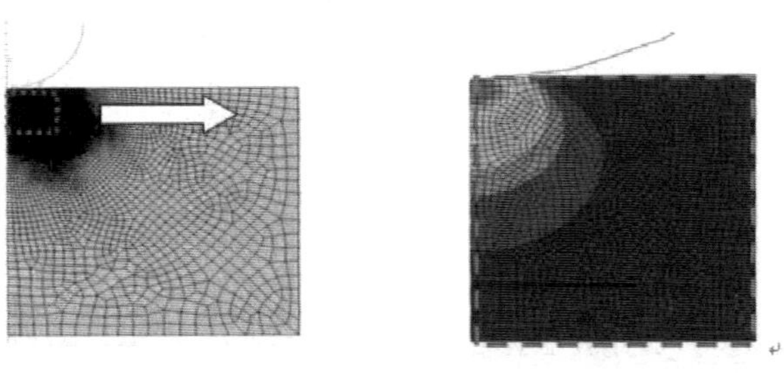

Figure 2. FE Model of the spherical indentation test (R=0.79 mm), (5b). Close up view showing the mesh underneath the indenter.

The bottom line of the model was fixed in all degree of freedoms (DOF) and the central line was symmetrically constrained. A gradient meshing scheme has been developed for different regions. While the spherical indenter models used R= 0.79 mm, and specimen model used young's modulus = 200 GPa, Poisson ratio = 0.2 and material plastic input data are to be used as input to the FE model include a carbon steel (0.1% C Steel) and mild steel. The mesh size is 10μm in the region underneath and around the indenter, while the mesh of other regions used single bias with bias element number 33 and ratio 5 to obtain gradient mesh tightly into underneath and around the indenter to improve the accuracy of the model. The free surface was made at the top and outside surface of specimen. In FE modelling, the accuracy of results is influenced by many factor such as the mesh shape and density, element type, friction condition and validation of the boundary conditions [6]. In the Spherical FE model the indenter considered as a rigid body to improve the modelling efficiency. A predefined displacement was applied on indenter and the reaction force is recorded on the reference point, representing the overall load on the indenter. The results of simulations FE model Spherical establish will produce p-h curve. Figure 6 shows typical P-h curve during loading and unloading phase of a typical elastic-plastic materials with different indenter sizes (R=0.50 mm; 0.79 mm; and 1.25 mm). The loading curve represents the resistance of material to indenter penetration, while difference between the loading and unloading curve represents the energy loss [7].The Spherical FE Model developed were validation with analytical solution of elastic material base of relationship using a known analytical solution [6] for indentation of linier elastic materials.

$$F_z = \left(\frac{16R}{9}\right)^{\frac{1}{2}} \cdot \frac{\mathrm{E}}{1-\mathrm{v}^2} \cdot \delta^{\frac{3}{2}} \tag{3}$$

Where ' F_z' is the reaction force, 'R' is the indenter radius; 'E' and 'v' is the Young's modulus and Poisson's ratio of material, respectively. 'δ' is the indentation depth. As shown in the figure 3a. visible the trend analysis in accordance with the numerical force–displacement data simulation FE Model and resulting using the following analytical. This indicates a statistically curve fitting the data equally well and the FE model is congruous with the analytical model. The correlation coefficients between these two curves using a least square regression method is within 99.9%. The FE spherical model was further validated by comparing the P-h curve with an elastic-plastic material model and published result data [8]. In the FE model, the material properties used were depicted directly from the published work. As shown in figure 3b the numerical results showed good agreement with the experimental data, which suggests that the FE model is valid and the results are accurate. One development in this work that has directly contribute to this process is the development of FE model. In this process firstly developed. Then the key parameter, such as dimensional data (e.g. sample size, indenter size) or material properties are identified and the parameters to be studied can be defined as variables. For example, in the initial model development process, the mesh size can be set as a variable to determine the best mesh density; the fictional coefficients at the indenter-sample interface can be set as a variable to determine the best values. In this stage, the main purpose is to find the P-h curve and material properties relationship, so the yield stress and the work hardening coefficients were set as variables. In this work encompassing a domain of Yielding strength σy from 100 to 900 MPa and strain hardening exponent n varying from 0.0 to 0.3 and Poisson's ratio v was fixed at 0.2.

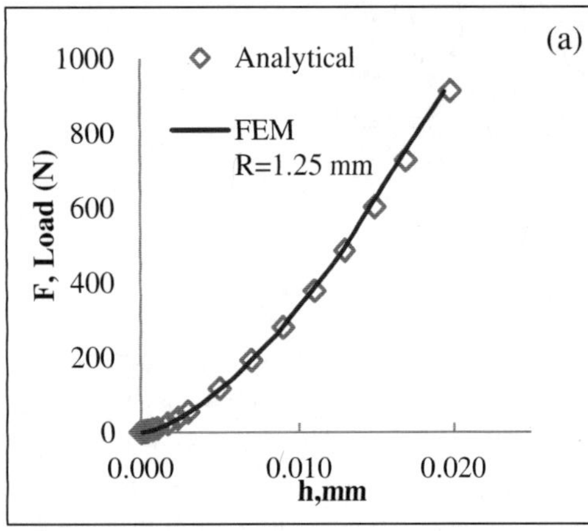

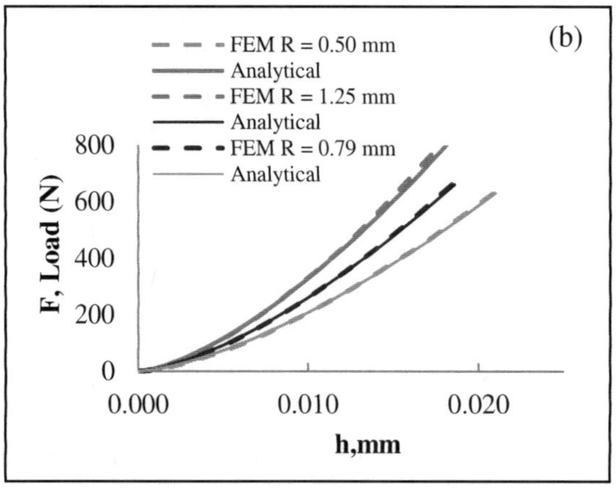

Figure 3a. Comparison between the FE numerical force–displacement (p-h curve) data and analytical solution with elastic material model. (R=0.5, 0.79, 1.25 mm.) and **Figure 3b.** Comparison of numerical results with published experimental data [8] of indentation with a spherical indenter (R=1.25 mm) showing the validation of FE model with elastic-plastic materials.

4. Effect of material properties and new analysing approaches of P-h curves

4.1. Full curve fitting approach and results

A Comprehensive parametric study using procedure developed was conducted representing the range of parameters of mechanical behavior found in common engineering metals. Poisson's ratio was fixed at 0.3, Young modulus E=200 GPa, the yield strength (σy) 100 to 900 MPa, and strain hardening from 0.0 to 0.3. The results of simulations with FE model Spherical establish will produce p-h curve. After evaluation of several approaches, two approaches have been found to be effective in representing the curves with adequate/acceptable accuracy. Figure 4-5 shows the two approached proposed. The first method is to use second order polynomial fitting in the form of:

$$P = C1.h2 + C2.h \tag{4}$$

The second fitting approach to be explored to represent the curve is using the force at different indentation depths. Most continuous machine can be depth controlled, and the representative stress is known to be depth dependent (angle dependent), so this potentially can be used as a more robust method. If the correlation between the force at different depth and the constitutive material properties and/or the representative stress is established, then the full P-h curve can be determined. This potentially can provide an effective way and physically meaningful way by using the power of computer simulation as large set of data has to be processed. For spherical indenter, the angle changes with the increasing depth, no fixed representative strain is readily available. However, in general, based on the deformation mechanism of an indentation process, the material deformation is controlled by the elastic deformation and the yielding, so we propose to use an effective representative stress which potentially could be linked to the C1 and C2 parameters, thus representing the full P-h curve. In the equation, C2 is linear term in the equation, so the fitting was conducted directly associating C2 to E/σr, this term represents the balance of elastic and plastic properties. Figure 4 plots the C2 vs. E/σr with different representative strain for C2. It clearly shows that there is a reasonable correlation between these C2 and E/σr, and the fitting is influenced by the representative strain used. At a representative strain of 0.01 (Fig.4), the fitting is reasonable with the best correlation coefficient.

$$C2 = 3566.9 (E/\sigma r) 0.855 \tag{5}$$

Theoretically, C1 is a second order coefficient, so it could potentially be linked to the representative stress follow the C1/σr vs E/σr according to Eq. 4 and Eq. 5, But the results is not very good with a effective representative strain of 0.07. Other strain level has been explored the results were equally not satisfactory. So a physical based hypothesis has been evaluated. From energy point of view, the resistance to indentation consists of elastic resistance and plastic resistance. The nonlinearity and linearity of the curve should reflect a balance between elastic deformation and plastic deformation, which can be represented by E/σr. The relationship directly between C1 and E/σr, has been explored with different representative strains. Some typical results are shown in figure 5. The fitting are much better than that for fitting between C1/ σ r vs E/ σ r. Comparing the correlation with different representative strains, the most effective reference strain is $\varepsilon r = 0.07$ (figure 5.(d)), which give an equation

$$C1 = 3606.8 (E/ \sigma r) (-1.252) \tag{6}$$

The correlation coefficients is over 93%. Further increasing or decreasing of the representative strains shows no improvement in correlation of the fitting. So this is the value in predicting the P-h curves to evaluate its accurate With the relationship between C1 and C2 with σr, the P-h curve can determined with a known set of material properties (yield stress and work hardening coefficients). Some typical examples are shown in figure-5. In each figure (a-c), the predicted P-h curves using the representative stress full curve fitting approach (solid line) and Finite element data (symbols) was plotted together. In each case, the comparison is in a reasonable agreement. The full curve fitting approach showed a reasonable accuracy in predicting the P-h curves with known material properties.

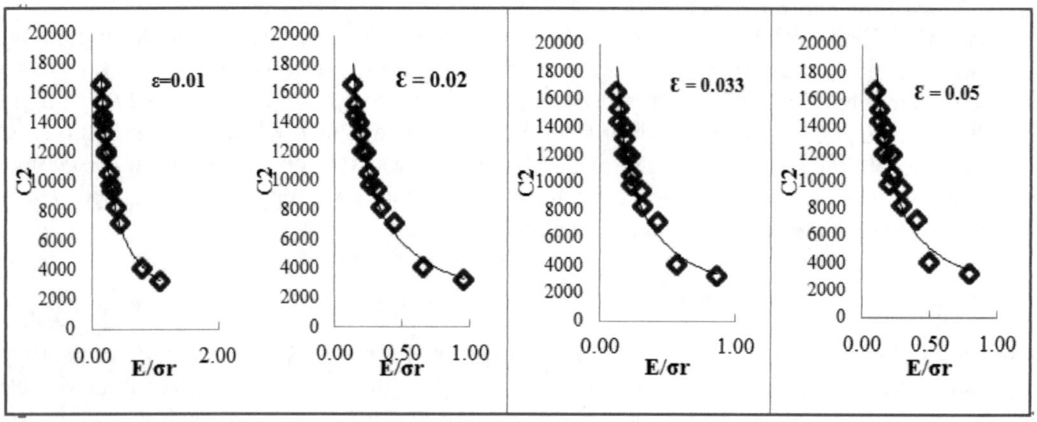

Figure 4. Plot the C_2 *vs.* E/σ_r with different representative strains for the full curve approach.

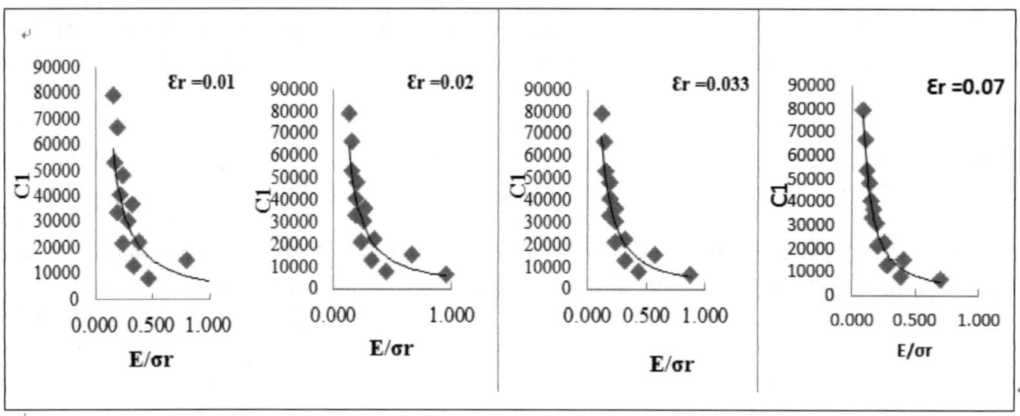

Figure 5. Plot the C1 vs. E/σr with different representative strains for the full curve approach.

4.2 Hardness prediction base on P-h curves an inverse material parameter estimation

The concept and methodology developed is to be use to predict hardness value (HRB) of material through direct analysis and validated with experimental data on selected sample of steel. Experimental test have been performed on 0.1% Carbon steel and Mild steel including tensile test and Rockwell hardness test (HRB). The Rockwell hardness is predicted directly base on P-h curve predicting using the representative stress equation established. **Figure 6a.** shows the P-h curves of the two steel samples from direct FE modelling (using the stress strain curves as input material properties) and the P-h curves predicted using equations (4-6). As shown in the figure, the curves are in a good agreement with each other. From the P-h curves, the indentation depth at a load of 100N and the 1000N is depicted and used to calculate the HRB following the equation:

$$HRB=130-(h_{1000N}-h_{100N})/0.002 \quad (7)$$

Figure 6b. plots the HRB values calculated based on the data in comparison with the experimental results. In all cases, the prediction based on the representative approach (equations 4-6) showed a good agreement with the test data and the FE data. Works has been conducted on other to materials, and the prediction results showed a similar degree of agreement. This suggests that the P-h curve based approach for predicting HRB values is valid and accurate.

Base on the concept prediction of hardness base on P-h curves an inverse material parameter estimation developed, it is implied that the hardness values could be potentially being directly linked to the representative stresses over the material ranges studied. Base on the concept prediction of hardness base on P-h curves an inverse material parameter estimation developed, it is implied that the hardness values could be potentially being directly linked to the representative stresses over the material ranges studied.
This is investigated by firstly determine the HRB values of materials over a wide range of properties, then a direct hardness-σ_r relation is explored using data fitting process. Figure 6 shows the correlation between the HRB *vs* σ_r

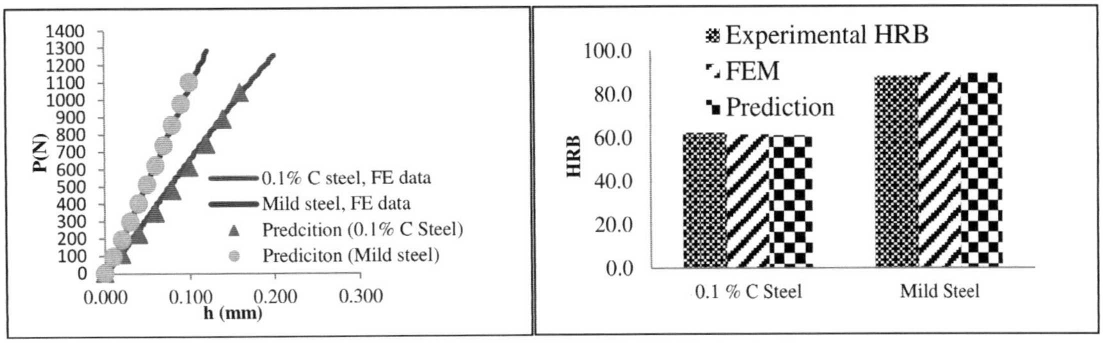

(a) (b)

Figure 6a. Comparison of the P-h curs of the two steel specimen from FE model and representative stress equations. **Figure 6b**. Comparison between experiment and prediction Rockwell hardness value (HRB) with FE modelling for Carbon Steel 0.10 % C and. Mild Steel.

$$HRB/\sigma_r = 0.0748 \ln (E/\sigma_r) - 0.2945 \quad (8)$$

These relationships (eqs.8) established allow direct hardness prediction from material properties. This is assessed using the two steel materials as example, the predicted HRB showed a similar level of agreement with the experimental data. In the case of the 0.1 C Steel, the HRB values is within 107% of the measured value; In the case of Mild steel, the HRB is within 102 % of the measured value. Similar agreement has been found in other materials (within 5% error range) as shown in figure 6(b). This suggests that these can be used to predict the hardness values with sufficient accuracy with the measurement error ranges.

5. Conclusions

In this work, the relationships between constitutive materials parameters (σ_y and n) of elasto-plastic materials, indentation P-h curves and hardness with spherical indenters has been systematically investigated by combining representative stress analysis and FE modelling using steel as a typical model material group. The main outcomes of work has formed a frame work of models to predict indentation P-h curves from constitutive
An approach to predict the P-h curves from constitutive material properties has been developed and evaluated based the relationship between the curvature and material properties and representative stress. Two new approaches to characterise the P-h curves of spherical indentation have been developed and evaluated. One is the full curve fitting approach while the other is depth based approach. In the full curve fitting approach, the relationship between an effective representative stress with the first and second order coefficients of a polynomial fitting line of the P-h was established. In the depth approach the relationship between force and representative stress with varying representative strain has been established. Both approaches were proven to be adequate/effective in predicting indentation P-h curves.

The approaches (i.e. predict hardness from P-h curves) established was successfully used to produce hardness values of a wide range of material properties, which is then used to establish the relationship between the hardness values (HRB) with representative stress.

6. References

[1] Fischer-Cripps A. C., 2007, Introduction to Contact Mechanics, Second Edition Edi, *Springer Science + Business media, LLC,* p 213

[2] I. N. Budiarsa, M. Jamal, "P-h Curves and Hardness Value Prediction for Spherical Indentation Based on the Representative Stress Approach", Applied Mechanics and Materials, Vol. 493, pp. 628-633, 2014

[3] N. Budiarsa, "Indentation Size Effect (ISE) of Vickers Hardness in Steels: Correlation with H/E", Applied Mechanics and Materials, Vol. 391, pp. 23-28, 2013

[4] Cao Y. P., Lu J., 2004, A new method to extract the plastic properties of metal materials from an instrumented spherical indentation loading curve, Acta Materialia, 52, 4023–4032

[5] Dao M., Chollacoop N., Van Vliet K. J., Venkatesh T. A. and Suresh S., 2001, Computational modelling of the forward and reverse problems in instrumented sharp indentation, *Acta Materialia*, Vol. 49, pp. 3899–3918

[6] Johnson K. L., 1985, *Contact Mechanics*, Cambridge: Cambridge University Press, UK,

[7] Swaddiwudhipong S., Tho K. K. , Liu Z. S. and Zeng K., 2005a, Material characterisation based on dual indenters, *International Journal of Solids and Structures*, Vol. 42, pp. 69-83.

[8] Kucharski S. and Mroz Z., 2001, Identification of plastic hardening parameters of metals from spherical indentation tests, *Materials Science and Engineering A,* Vol. 318, pp. 65-76